W9-BCU-094

CLARENDON LIBRARY OF LOGIC AND PHILOSOPHY
General Editor: L. Jonathan Cohen, The Queen's College, Oxford

PHILOSOPHY WITHOUT AMBIGUITY

The Clarendon Library of Logic and Philosophy brings together books, by new as well as by established authors, that combine originality of theme with rigour of statement. Its aim is to encourage new research of a professional standard into problems that are of current or perennial interest.

General Editor: L. Jonathan Cohen, The Queen's College, Oxford.

Also published in this series

Quality and Concept by George Bealer
Psychological Models and Neural Mechanisms by Austen Clark
The Probable and the Provable by L. Jonathan Cohen
The Diversity of Moral Thinking by Neil Cooper
The Metaphysics of Modality by Graeme Forbes
Interests and Rights: The Case Against Animals by R. G. Frey
The Logic of Aspect: An Axiomatic Approach by Antony Galton
Ontological Economy by Dale Gottlieb
Equality, Liberty, and Perfectionism by Vinit Haksar
Experiences: An Inquiry into some Ambiguities by J. M. Hinton
The Fortunes of Inquiry by N. Jardine
Metaphor: Its Cognitive Force and Linguistic Structure by Eva Feder Kittay
Metaphysics and the Mind–Body Problem by Michael E. Levin
The Cement of the Universe: A Study of Causation by J. L. Mackie
Divine Commands and Moral Requirements by P. L. Quinn
Reason by Nicholas Rescher
Simplicity by Elliott Sober
The Logic of Natural Language by Fred Sommers
Blindspots by Roy A. Sorensen
The Coherence of Theism by Richard Swinburne
Anti-Realism and Logic by Neil Tennant
The Emergence of Norms by Edna Ullmann-Margalit
Ignorance: A Case for Scepticism by Peter Unger
The Scientific Image by Bas C. van Fraassen
The Matter of Minds by Zeno Vendler
Chance and Structure by John Vickers
What is Existence? by C. J. F. Williams
Works and Worlds of Art by Nicholas Wolterstorff

Philosophy Without Ambiguity

A Logico-Linguistic Essay

JAY DAVID ATLAS

CLARENDON PRESS · OXFORD
1989

Oxford University Press, Walton Street, Oxford OX2 6DP
Oxford New York Toronto
Delhi Bombay Calcutta Madras Karachi
Petaling Jaya Singapore Hong Kong Tokyo
Nairobi Dar es Salaam Cape Town
Melbourne Auckland
and associated companies in
Berlin Ibadan

Oxford is a trade mark of Oxford University Press

Published in the United States
by Oxford University Press, New York

British Library Cataloguing in Publication Data
Atlas, Jay David.
Philosophy without ambiguity: a logico-linguistic essay.—(Clarendon library of logic and philosophy).
1. Linguistic philosophy
I. Title
149'.94
ISBN 0–19–824454–1

Library of Congress Cataloging in Publication Data
Atlas, Jay David.
Philosophy without ambiguity: a logico-linguistic essay / Jay David Atlas.
p. cm.—(Clarendon library of logic and philosophy)
Bibliography: p. Includes index.
1. Language and logic. 2. Semantics (Philosophy). 3. Pragmatics. 4. Ambiguity. I. Title. II. Series.
BC57.A85 1989 149'.94—dc19 88–8017 CIP
ISBN 0–19–824454–1

Set by Hope Services, Abingdon
Printed in Great Britain by
Biddles Ltd., Guildford and King's Lynn

To Jacob Henry Atlas, Babette Friedman Atlas
and
the memory of Bluma Cohen Atlas

Preface

My colleagues have sometimes asked me how I came to have such (how shall I put it?) idiosyncratic views; when I first suggested some of the ideas discussed in this book, in Atlas (1974, etc.), some of my philosophical colleagues were more than a little sceptical. A few of them were downright annoyed. My family put it down to my being an Atlas, of the tribe of Abarbanel. There may be other, better, sociological answers. In any case, there is a bit of history to this, and for what light it can shed on an intellectual research project I shall briefly describe it here.

The aeroplane descended for its landing at Detroit Airport, and my eyes were ravished by the deep green of the stand of trees at the edges of the runway, a green of summer rain, a green whose richness excited in me a feeling of joy that I had not felt in the months since I had left Princeton. It was June 1973. I was twenty-eight, and at the invitation of George Lakoff and Lauri Karttunen I was spending my summer as a Fellow of the Workshop on the Formal Pragmatics of Natural Language in the University of Michigan, Ann Arbor. Joining me in the Workshop, among others, were Stephen C. Levinson, Jerry Sadock, and Larry Horn, who were to play greater roles in my intellectual life than I could ever have anticipated that rainy afternoon. As I walked down the steps from the aircraft to the runway I felt warm raindrops cascade down my nose and ears, and down my neck, while thunderclaps rolled away into the distance.

Two and half years before, during a spring in Princeton and a summer in Oxford, with Dana Scott, I had begun to connect my studies of Scott–Montague intensional logic, Generative Semantics, and Donald Davidson's philosophy of language with problems of Dummett, Strawson, and Russell. I was nourished by frequent conversations with Dana Scott and Michael Dummett during those months.

In Michigan I was resuming these studies in a new setting,

reading Dummett's newly published volume on Frege's philosophy of language, lecturing on Frege's account of presupposition, and, under the influence of George Lakoff and Stephen Levinson, trying to understand why linguists were preoccupied with Paul Grice's 1967 William James Lectures at Harvard University. I was grateful to Paul Grice, for he had himself brought me a rare copy of "Logic and Conversation" during a visit he had paid to Princeton, but I had not been able to do anything with it. I was immersed in problems of truth, logical form, and formal semantics originating in the work of my teachers Michael Dummett, Donald Davidson, Gilbert Harman, Richard Grandy, Paul Benacerraf, and Dana Scott.

Thinking again about Dummett's and Strawson's discussions of presupposition and truth, and about van Fraassen's theory of supervaluations, walking and talking in the quiet twilight with Stephen Levinson amid the Gothic-Revival halls of the University of Michigan Law School, in July 1973, I had a happy thought that has since led me to investigate the boundary between the semantics and the pragmatics of language: (1) *Natural language negation could not be ambiguous*. Van Fraassen and Łukasiewicz were wrong; Russell and the published Strawson were both wrong; Dummett and Waismann, and, it turned out, the unpublished Strawson as well, had worried about the right issue; and suddenly Grice's William James Lectures became important to me.

Some weeks later, during the summer meeting of the Linguistic Society of America in Ann Arbor, I walked into the giant auditorium of the Modern Language Building, where Jerry Sadock was giving a paper of his and Arnold Zwicky's on ambiguity. He was in the middle of the paper and, since I could not pick up the threads of the argument, I left the hall, making a mental note to ask him later for a copy. It was in July 1974, at Hal, Annabelle, and Mark Jensen's house in Downingtown, Pennsylvania, that I began to read the Sadock and Zwicky paper. Downingtown is one stop beyond Paoli on the Main Line. Its gently rolling hills, whose features I had seen and felt at every season, had cradled my recent intellectual life. In this sunny, fertile, tactful Pennsylvania landscape I read the first paragraph of Zwicky and Sadock's paper and suddenly reached the second part of one of the more interesting hypotheses of my intellectual life: 'not' was not ambiguous, (2) because *it was general in sense* among exclusion and choice-negation interpretations, among contradictions and contraries, among wide-scope and narrow-scope interpretations, and even, I suspected, among object-language and metalanguage uses.

By the time I had finished the first paragraph of Zwicky and Sadock's paper, these conclusions about negation had become a moral certainty to me. I quickly discovered that Zwicky and Sadock had not discussed negation in their paper, so I set to work immediately to test my hypothesis using their ambiguity tests. The tests confirmed, to my mind, the generality-of-sense hypothesis, and within the hour I knew that I had an Archimedean point from which to reconstruct entirely my thinking about logical form, theory of truth for natural language, presupposition, the Russell–Strawson debate on definite descriptions, Donald Davidson's account of the logical form of action sentences, the semantics of adverbs, and Gricean implicature, among other central topics in the logic of grammar.

G. Lakoff's ideas on "performative antinomies", in Lakoff (1972), had led to my discovering "assertoric antinomies" (Atlas (1972), discussed in G. Lakoff (1975: 266–7)), which focused my interest on *the interface between semantics and pragmatics*. A major portion of my work since 1973, and in particular the research discussed in this book, has developed a theory of that interface. That theory can be encapsulated in a Kantian slogan, to be elaborated in this book: *Pragmatic inference without sense-generality is blind; sense-generality without pragmatic inference is empty*. (My philosophical readers, but perhaps not all my linguistic readers, will know Kant's famous apophthegm from the *Critique of Pure Reason*: 'Concepts without percepts are empty; percepts without concepts are blind'.) This account, first suggested in Atlas (1975*b*) and developed in Atlas (1977*b*, 1979), seemed to me required by my discovery of the philosophical importance of generality of sense and the semantic contrast between the generality of sense and ambiguity. That contrast and its consequences are major topics of this book. Once again the moral I drew, in Atlas (1974, etc.) can be encapsulated in a slogan, the thesis that *the generality of sense radically underdetermines truth-conditional content*, or, to put it more pithily in philosophical jargon, the sense of a sense-general sentence is not a proposition, the bearer of determinate truth or falsity. A correlative philosophical thesis was that *sense-general sentences do not have Russell–Tarski* *logical forms*. Much of this book is devoted to an updated explanation and defence of these theses. In the course of my work, 1973 to the present, as I considered more linguistic data and refined my theory, I naturally reformulated my views. This book provides a reconsidered and canonical version. A comment of Larry Horn's (1985: 154) persuaded me to do this.

As an undergraduate, taught by Joe Epstein at Amherst College, reading Ernest Nagel, Morton White, and W. V. O. Quine, and studying mathematics, I accepted extensionalism, behaviourism, and pragmatism as if they were mother's milk. Then I was taught by Joe Swanson, Richard Grandy, and Michael Dummett, and supervised first by Dana Scott and then by Paul Benacerraf. They led me to rethink intensionalism, Chomskyan mentalism, and Fregean–Russellian realism, and to read in the philosophy of mathematics and the philosophy of language. I spent a lot of time reading every chapter but chapter two in Quine's *Word and Object*. Chomsky's *Language and Mind* and his (1959) review of Skinner's *Verbal Behavior* were a liberation from behaviourism. Dana Scott's instruction in intensional logic and Paul Benacerraf's iconoclasm were a liberation from extensionalism. And thanks to the writings of Dummett, Strawson, Cohen, Katz, Kripke, and Chomsky I began to take meaning, not just reference, seriously. There are and have been plenty of people discussing reference, speech acts, truth, and logical form. Not too many people in the philosophical tradition in which I was educated were taking linguistic meaning seriously, e.g. synonymy, analyticity, category mistakes (selection restrictions), etc. There is an interesting history as to why not. The views I have arrived at are not, then, obvious developments of the views in which I was educated; but I could not have got here without them.

J.D.A.

Princeton, New Jersey

Acknowledgements

As this has been a long-term research project, a goodly number of friends and colleagues have advised me in earlier stages of this work. I am grateful to Rogers Albritton, Jens Allwood, Jacob H. Atlas, Kent Bach, Morton Beckner, Paul Benacerraf, Sir Isaiah Berlin, David Braun, Penny Brown, Tyler Burge, John Burgess, Susan Castagnetto, Jonathan Cohen, Hugh Collins, Rob Content, René Coppieters, Keith Donnellan, Michael Dummett, the late John Paul Egan III, Joseph Epstein, Robert Evrén, Gerald Gazdar, Michael Geis, Edmund Gettier, David Glidden, Richard Grandy, the late Paul Grice, Jess Gropen, Howard Gruber, David Harrah, Larry Horn, Dan Isaacson, Mark Jensen, Philip Johnson Laird, Will Jones, Lauri Karttunen, Jerry Katz, Ruth Kempson, David Keyt, Thomas S. Kuhn, George Lakoff, Stephen Levinson, William Lycan, Ron Macaulay, Brian Maddox, John N. Martin, Al Martinich, James McCawley, Adam Morton, Larry Roberts, Michael Rosen, David Sachs, Jerry Sadock, Henry Schankula, Dana Scott, Pieter Seuren, Mark Shepherd, Robert Sleigh, Scott Soames, Kurt St Angelo, Keith Stenning, Sir Peter Strawson, Andy Strominger, Rick Tibbetts, Todd Tibbetts, Charles R. V. Tomson, Martin Tweedale, Frank Veltman, Wang Hao, Lucia P. White, Morton G. White, Stephen D. White, and Deirdre Wilson.

I owe a special debt of gratitude to Robert A. Dennis, James S. Dittmar, Mikael Dolfe, Sean Diarmuid Fennessy, Geoffrey Granter Koziol, William Leach, Charles Peter Olson, and Mark Allen Phillips.

This book was conceived and written at the Institute for Advanced Study, Princeton, New Jersey, during my research there in 1982, 1983, 1984, and 1986. For their hospitality, generosity, and intellectual advice during my Research Associateship at the Institute, I am particularly grateful to Lucia and Morton G. White. This book would not have been written without them. I am also grateful to Homer Thompson and Marshall Clagett. Enid Bayan, Sabina Modzelewski, and Paul Schuchman assisted me in many ways, as did Joyce Cokinos, Charles Greb, Glenda Whitmore, and Mary Wisnovsky. I also thank Peggy Adams, Madalene Bottom, Pearl Cavanaugh, Ann Getson, Anne Johnson, Arlene Lundergan, and Dot Rand.

Linguists, psychologists, and philosophers in various universities have heard versions of this work, and I have benefited from their comments. I thank my colleagues in Amherst College, University of California at Los Angeles, at Riverside, and at San Diego, University of Cambridge, University of Chicago, University of Göteborg, University of Hong Kong, University of Kansas, University of London, Pomona College, Princeton University, Rice University, Rutgers University, University of Southern California, Stanford University, University of Stockholm, University of Sussex, Wolfson College, Oxford, and the Institute for Advanced Study, Princeton, New Jersey.

For permission to reprint excerpts from published work, I thank the Institute of Electrical and Electronic Engineers Computer Society Press, Academic Press, Blackwell, D. Reidel, and the editors of *Linguistics and Philosophy*, *Mind*, and *Philosophical Books*. For editorial advice I am grateful to Angela Blackburn, Hilary McGlynn, Molly Scott, and Jane Wheare.

Finally, among those who have taught me much, I wish particularly to acknowledge Paul Benacerraf, Larry Horn, Ruth Kempson, George Lakoff, Stephen Levinson, Jerry Sadock, and Dana Scott, upon whose intellect, imagination, and friendship my intellectual work has so heavily relied, and John Francis Walter, playwright and poet, whose untameable energy and restless curiosity have made life worth thinking about. The dedication of this book to my father, mother, and the memory of my paternal grandmother speaks for itself. They have all helped make the house of intellect into a home.

Contents

A Note on Notation

I use notation and typography from logic, mathematics, and linguistics in this book. There are also the usual literary conventions in the use of quotation marks to deal with. As a reader's guide, I briefly sketch my usage here.

1. Double Quotation Marks
   (*a*) "Snow is white" names the sense of the sentence-type a token of which is here exhibited between the double quotation marks. (I could have used Wilfrid Sellars's (1963: 204–5) dot quotation, e.g. •Snow is white•.)
   (*b*) I use double quotation marks in direct literary quotation when I am using the word-types of others.
   (*c*) I use double quotation marks when I scare-quote.
2. Single Quotation Marks
   (*a*) 'Snow is white' names an English sentence-type, a token of which is here exhibited between the single quotation marks.
   (*b*) Single quotation is also used for direct quotation within a direct quotation.
3. Italics
   (*a*) *Snow is white* names an utterance-type, a token of which is here italicized.
   (*b*) Italics are used for rhetorical emphasis.
   (*c*) I italicize some words, which I use, that are borrowed from a non-English language.
4. Small-capital Letters
   Small-capital letters indicate contrastive stress, for emphasis, in utterance-types.

If here we take a bold-face expression to name the very token so written, a nonce name for itself each time the token occurs, we can state:

(*a*) **John Walter** is an utterance-token of the utterance-type *John Walter* and of the name-type 'John Walter'.
(*b*) **John Walter** is a *John Walter*.
(*c*) **John Walter** is a 'John Walter'.
(*d*) 'Snow is white' means "Snow is white".
(*e*) 'Snow is white' means that snow is white.
(*f*) **SNOW is white** is not a token of *Snow is white*, or of *Snow is WHITE*, but is a token of 'Snow is white'.

The interpretation of mathematical and other notations should be obvious from context.

Sometimes . . . philosophy is no more
than semantic inertia; guarding
how we should speak and think
through precepts of nagging custom.

(Richard L. Gregory)

Science is not "organized common sense";
at its most exciting, it reformulates
our view of the world by imposing
powerful theories against
the ancient, anthropocentric
prejudices that we call intuition.

(Stephen Jay Gould)

Introduction

This essay discusses the results of a research programme in theoretical linguistics, philosophy of logic, and philosophy of language that has engaged my attention since 1973. It is a programme that, fortuitously, has provided philosophical analyses and theoretical foundations for some recent research in linguistics, philosophy, and psychology—e.g. Deutsch (1987), Deutsch and Feroe (1981), Bach (1982), Lewis (1982), Gazdar (1979, 1980), Horn (1985, 1988, 1989), Kempson (1986, 1988), Levinson (1985, 1987), and Sperber and Wilson (1986).

The focus of this work is the following question: When one hears a sentence spoken in a context, knows its literal meaning, and understands what the speaker's utterance meant, what is the connection between one's knowledge of the literal meaning of the sentence and one's understanding of the utterance-meaning of the statement made?

This is a very general question, and I do not intend to deal with it as generally as I might. That kind of generality is a philosophical penchant, and I intend to resist my temptation to indulge it. I mean to approach the problem piecemeal, focusing on a logically crucial, scientifically manageable case: the case of ambiguous versus univocal, semantically general, sentences. When one hears an *ambiguous* sentence spoken in a context, knows its various literal meanings, and understands what the speaker's utterance meant, one connection between knowledge of sentence-meaning and utterance-understanding is this: taking account of the context, one *selects* that *sense* of the sentence that best fits the context. Since tense, indexicals, etc. must also be evaluated in context, selection of a sense does not complete one's understanding. But, of course, it is essential to it.

The relatively common-sense description that I have just given appeals to a number of theoretical notions: ambiguity, context, literal meaning, sentences, statements, sentence-meaning, utterance-meaning, best fit. Some notions from Gricean pragmatics and speech-act theory I shall merely use and not explore, e.g. context, statement, utterance meaning. My ideal reader is (or can become) acquainted with works like Atlas (1984*a*), Atlas and Levinson (1981), Bach and Harnish (1979), Grice (1975), Horn (1984*b*, 1988), Leech (1983), Levinson (1983), and Sperber and Wilson (1986). Some notions from linguistic semantics and from formal semantics I shall both use and explore, e.g. ambiguity, semantic representation, literal meaning, logical form. But I hope that my ideal reader is (or can become) acquainted with works like Cohen (1971), Cruse (1986), Davidson and Harman (1972, 1975), Dowty *et al*. (1981), J. D. Fodor (1977), Johnson (1987), Katz (1972, 1986, 1987), Kempson (1977), Lakoff (1986, 1987), LePore (1987), Lycan (1984), Lyons (1977), Martin (1987), and van Fraassen (1971). I do not assume knowledge of elementary Chomskyan syntax, though there is some syntactic argumentation in this book. For readers entirely new to Chomsky's work, Chomsky (1986), Radford (1981), and Riemsdijk and Williams (1986) will give the flavour of it.

The book leaves aside many theoretical complexities and relies on the reader's good sense, linguistic intuition, and attentiveness to his or her "ear" for language. There is more than enough complexity in the notions that are my primary concern: ambiguity, generality of sense, presupposition, negation, topic/comment in statements, and conversational inference, e.g. Grice (1975) implicatures. These topics provide the foci of this book. I review the Atlas-Kempson theory of negation and my resolution of the Russell–Strawson debate on presupposition, definite description, and truth-value gaps, give a new treatment of negative existence statements (with a solution to an anomaly in Strawson's views, a correction of Quine's account of Russell in Quine's "On What There Is", and a demonstration that Russell's problem of existence statements was a linguistic pseudo-problem), and then draw some general lessons from the resolution of these problems for the theory of meaning in linguistics and philosophy. In particular, I argue that philosophers have misapplied the notion of ambiguity. They have often said that proper names are "ambiguous", or that the referential and attributive uses of definite descriptions amount to an ambiguity, a view which, despite being defended by Barwise and Perry (1983), was properly criticized by Kripke (1979). But these are the obvious cases. This book is about the unobvious cases, e.g. the alleged scope ambiguity of negation. (I have doubts about alleged quantifier-scope ambiguities and about alleged transparent, translucent, and opaque "readings" of propositional attitude sentences, but I do not present my views here; see Bach (1982), Kempson and Cormack (1981), and Sadock (1975).)

If sentences heretofore thought to be ambiguous are not, but are univocal, semantically general, instead, then the apparent "readings" or "senses" are really "complete" pragmatic interpretations derived in part from semantically "incomplete" senses. The picture of the connection between knowledge of sentence-meaning and utterance-understanding alters. When one hears a *sense-general* (not *ambiguous*) sentence spoken in a context, knows its literal meaning, and understands what the speaker's utterance meant, one connection between the knowledge of sentence-meaning and utterance-understanding is now this: taking account of the context, one *constructs* (not *selects*) that *interpretation* (not *sense*) of the sentence that best fits the context.

This book discusses the notion of generality of sense, contrasts it with ambiguity, and considers the consequences of sense-generality for our theories of logical form and meaning, of negation and presupposition, of the interface between semantics (meaning) and pragmatics (use). In its emphasis on univocality, it follows the lead of G. E. M. Anscombe, W. V. O. Quine, and, especially, Morton G. White. I hope that the reader will find that the perspective offered by the book on both classic and new problems in the philosophy of language will invigorate the subject.

Linguists and philosophers find it natural to split their analyses of language into three levels: the sentence (grammar), the statement (speech-act theory), the speaker (pragmatics). There is a very strong tendency for logicians and syntacticians to focus on the sentence, and for philosophers to want to reduce statements to speakers' meanings. I inveighed against this in Atlas (1975*a*), when I argued that Frege had three notions of presupposition, one at each level. The sentence may be thought of as an abstract object, e.g. as in Katz (1981), and studied mathematically. The speaker may be thought of as a mental entity and studied psychologically. What, then, has happened to that convention-bound, truth-bearing, intention-laden stuff that is *parole*? The Grice (1975) notion of a *generalized* conversational implicature takes the intermediate level of theory seriously, flanked on the one hand by truth-conditions ("what is said") and on the other hand by context-bound, *particularized* conversational implicature (what a speaker implicates by an utterance in a particular context). The reader will discover that I take all three levels seriously.

In this book I introduce the reader to visual, semantical generality in Chapter 1, where I give a novel and accurate interpretation of Locke's famous 'triangle' paragraph on Abstract Ideas, modify the psychologists' standard accounts of visual illusions, discuss Gombrich on the psychology of art, and contrast sense generality with ambiguity. In Chapter 2 I get down to linguistic work, introducing verbal ambiguities and stating the Fundamental Theses of my view in Section 0. In Section 1 I discuss criteria of ambiguity offered by Quine (1960) and Lyons (1977). In Section 2 I elaborate the criteria and discuss parts of the classic paper on ambiguity by Arnold Zwicky and Jerry Sadock (1975). This significant paper should not be ignored by philosophers, logicians, psychologists, computer scientists, or linguists. In Section 3 I try to head off a natural misconstrual of generality of sense. Thus Chapter 2 is a systematic, linguistic discussion of the difference between ambiguity and generality of sense. In Chapter 3 I apply Chomskyan linguistics to several famous problems in philosophical logic. In particular I apply the Zwicky–Sadock ambiguity tests to negative sentences. They fail. The classic view that negation is scope-ambiguous is false. I then consider the consequences of this observation for Bertrand Russell's and Peter Strawson's theories of sentences containing definite descriptions, for truth-value gaps, and for logical forms as representations of the sense of *sentences*. I argue that logical forms cannot be the semantic representations of sense-general sentences. I then develop Strawson's views on the topic/comment structure of *statements* and give novel solutions to the different Russellian and Strawsonian problems with negative existence statements. If my arguments are correct, the Russellian problem of negative existence statements is a linguistic pseudo-problem, and the Strawsonian problem has an elegant solution. Along the way I provide an account of the history of Russell's philosophy in the remarkable years 1903–5. Finally, in the last section of Chapter 3 I show that the doctrine of the generality of sense of negative sentences solves an ancient anomaly in the logical tradition's account of negation. I find this solution particularly satisfying. In Chapter 4 I consider what lessons I can draw for the theory of meaning from the linguistic study of generality of sense and the new analysis of negation, presupposition, and existence statements. In particular I address the questions of the adequacy of a Tarski–Davidson theory of truth for natural language and of possible-world semantics, the relation between sentence-meaning and utterance-understanding—e.g. the role of generality of sense in explaining the understanding of utterances by Gricean pragmatic inference—and the consequences of the generality of sense for a semantics of Natural-Kind terms. The conclusions I reach are novel and have implications for linguistic theory (see Kempson: 1988).

So much by way of background and orientation. The philosophy, logic, syntax, semantics, and pragmatics in this book are all, by most academic as well as lay standards, somewhat technical, but it is not Real Analysis or Quantum Mechanics. Behind the technical discussion in this book there slowly develops a general "picture", in the later Wittgenstein's sense, of language. I hope that it is a correct picture, but whether it is or not, I hope that it provokes, possibly excites, others to do something better or different.

1

Abstract Ideas, Abstract Art, and Abstract Language

In our language General terms, i.e. those with divided reference like the common noun 'apple', contrasted with Singular terms, like the definite description 'the President of the United States', are *general in sense*, i.e. their meanings are *nonspecific with respect to certain predicates*, some of which are "basic" in a lexicon of English. For example, to say that an object is an apple is not by virtue of the meaning of 'apple' to attribute a *size* to the object, unlike 'dwarf', or a *colour*, unlike 'palomino'. (Redness is only stereotypical for apples; 'pumpkin-sized green apple' is not an oxymoron.)[1] In contrast, images, especially mental ones, like Bishop Berkeley's Ideas, have been supposed to be quite specific. The mental image that I conjure up when I imagine a speckled chicken has been taken to be an image of a chicken with eighty-seven speckles, or one with eighty-eight speckles, etc. Nevertheless, it seems to me that I can have a perfectly good image of a speckled chicken without my image being specified for the number of speckles. As Paul Ziff (1972: 136) once wrote, "Staring at a pictorial representation of a man in ordinary black opaque unbulgy riding boots we need not ask: is that man a web-footed or a nonweb-footed fellow? A real live man must be one or the other, but a man in a picture is not a real live one." The same goes for feathered bipeds as well.

Still, there are those who would save something of the Bishop's

[1] These "basic" predicates are those from the lexicon of any successful grammar that would be employed to state the meanings of English words, especially the contrasts among those lexemes that are members of "contrast sets", e.g. [MALE]/[FEMALE] for {'mother', 'father'}, {'mare', 'stallion'}, etc. (see Lehrer (1974), Lyons (1977), Pulman (1983), Grandy (1987: 261, 272)). Examples may be found in decompositional theories of meaning proposed by J. J. Katz (1972), G. Lakoff (1971), and G. Miller and P. N. Johnson-Laird (1976). I shall not assume the correctness of any kind of lexical decomposition theory for lexemes of English. Some of the difficulties with them have been interestingly discussed by Pulman (1983) and Grandy (1987). I would be content if "basic" were a psycholinguistic predicate in a theory of language acquisition. See G. Lakoff (1987).

doctrines. John Mackie (1976: 124) wrote that "the generality of ideas can be far more extensive than the indeterminacy of images, and it is reasonable to suppose that it arises from a quite different source". Jerry Fodor (Fodor 1983*a*: 98) has claimed that if:

> you look at the case of language . . . it looks like the kind of representational capacity you need, the kind of information you have to represent, the kind of structures that you have to be able to display and the operations you have to be able to perform upon them, are really very highly specific to language. The analogies between, say, language and visual perception seem to be really pretty weak and unimpressive. So it might be that, not only are there these specialized computational systems, but that they, as it were, talk individual private languages; that the representational systems you use for doing visual form perception are really quite different from the ones that you use for doing, say, the syntactical analysis of utterances.

As a matter of human physiology, even if I knew what it meant to "do" visual-form perception, I do not know what to compare in the "representational systems" I use for understanding what I see and for understanding the sentence-tokens that I hear. For one thing, I do not know what those systems are. I think I can compare how I go about "seeing" pictures, maps, paintings, and diagrams with how I go about "understanding" word- or sentence-types and word- or sentence-tokens. I can compare features of the representations themselves, and I can compare the various relations "*x* represents *y*", "*x* represents *y* as *z*", or "*x* represents *y* as *z* to observer *o*" that arise from various values of '*x*', '*y*', '*z*', and '*o*'. These sorts of comparisons seem within my commonsensical and scientific ken. I certainly do not think that mental images are exactly like painted or drawn ones (a safely trivial remark). I shall be concerned primarily with non-mental representations, their properties and relations, though I shall speculate about states of mind too, including the having of mental images.

For example, it is natural to suppose that the sense of 'apple' does not specify a size or a colour, but that one's mental image of an apple possesses a specific size and colour. In Mackie's view the general notion of an animal covers giraffes, earthworms, and oysters, and he (1976: 124) wrote that "no one could notice an animal while remaining completely vague about whether it was an oyster, an earthworm, or a giraffe". Nevertheless, I think that I could notice an animal, and notice that something was an animal, while remaining quite vague about what sort of animal it was. For example, I see a flash of pelt as a marmoset runs through the ferns. For all I can tell, the flash of the pelt and the waving of ferns may have been caused by a giraffe in my garden, buried, in this case, not up to his neck but up to his head in my fern-bed. Or suppose Professor Robert Nozick stands in front of me with a bag over his head. Couldn't I notice a human being while remaining completely vague about whether it was a Harvard philosopher?

It seems to me, unlike Mackie (1976) and Bennett (1971: 42–3), that nothing in the grammar of 'notice', in the nature of perception, or in the character of physical objects excludes this vagueness. Furthermore, even if the generality of concepts were "more extensive" than the indeterminacy of images, nothing follows as to the difference of the representational characters of concepts and images. Mackie and Fodor offer no grounds for ruling out the similarity of the representational characters of General terms and indeterminate images. *They certainly have given no reason to deny that our cognitive abilities to understand General terms and to recognize indeterminate images are to be explained by similar neurophysiological or psychological concepts.* Prima facie the more intellectually interesting hypothesis is that there are some identities of neurological function, and perhaps structure, underlying our faculties to see, hear, and understand what we see and hear. It seems to me that whatever modularity the brain turns out to have will not affect this point.[2]

(A good example of a modularity approach is found in Lightfoot (1982). Lightfoot (1982: 45) writes:

> The grammar is kept simple and a separate perceptual strategy is exploited, which assigns parallel interpretations to adjacent ambiguous objects, as suggested by studies of vision: a Necker cube is ambiguous in terms of which square surface is nearer to the observer, but if two are placed side by side, they will usually be interpreted in the same way unless the observer engages in some visual gymnastics. If we interpret this perceptual strategy generally so that it applies not only to visual objects like Necker cubes and the like but also in the cognitive sphere, then the grammar may have a simple rule allowing a pronoun to refer back to any

[2] For a suggestion that it is reasonable to suppose that the generality of concepts and the indeterminacy of images have the same source, see G. Lakoff (1977). For a discussion of modularity, see Fodor (1983*b*), Gazzaniga (1985), and Ornstein (1986).

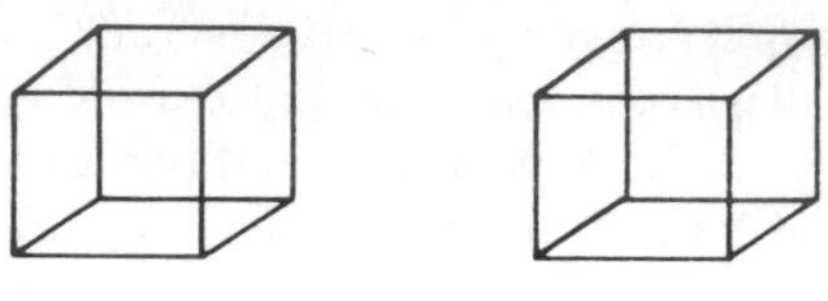

Fig. 0

noun phrase on its left. This grammar allows *Brian persuaded Bill to fix his bicycle and John persuaded Jay to fix his car* to have four possible logical forms, but only two of them could actually be perceived by the mind. Similarly for *John must do the shopping and Brian must too*. In this way a simple and general rule of grammar intersects with a simple and general perceptual strategy, and in combination they yield the correct predictions: namely that *Brian persuaded Bill to fix his bicycle and John persuaded Jay to fix his car* has only two relevant interpretations. This illustrates the modular approach to the study of mind: the mind is reckoned to consist of distinct subsystems, which interact in specified ways.)

I conclude that our ability to understand General terms when we hear or see tokens of them might well require some version of John Locke's "abstract Ideas", not because the meaning of a sense-general word must be determined by an Idea or image in the mind that lacks certain features, as Locke might have thought, but because our ability to understand what words mean might be constrained by the same biological mechanism and interpretative methods that constrain our ability to cognize what images display. In any case Locke's views have been misunderstood by philosophers who ignore the Scholastic semantic tradition in which he was educated. Recall that Locke (*Essay* III. iii. 7–8) writes:

There is nothing more evident than that the ideas of the persons children converse with are, like the persons themselves, only particular. The ideas of the nurse and the mother are well framed in their minds; and LIKE [*my emphasis*] pictures of them there, represent only those individuals; and the names of *nurse* and *mama* the child uses, determine themselves to those persons. Afterwards, when time and a larger acquaintance has made them observe that there are a great many other things in the world, that, in some common agreements of shape and several other qualities, resemble their father and mother and those persons they have been used to, they frame an idea which they find those many particulars do partake in; and to that they give, with others, the name *man*, for example. And thus they come to have a general name, and a general idea. Wherein they make nothing new, but only leave out of the complex idea they had of

Peter and James, Mary and Jane, that which is peculiar to each, and retain only what is common to them all.

'Man' is a semantically general term that is in its sense unspecified for gender, unlike the anachronistic 'poetess'; hence it is referentially unspecified for the sex of its denotation. One should note the difference between *sense and reference notions of generality*. Locke notes that the idea is abstracted from ideas of Mary and Jane as well as of Peter and James. On Locke's view no one who understood seventeenth-century English could take a term *sense-unspecified for gender* to be *referentially specific of the male sex*, i.e. to be true of only males of the species.[3] He continues:

By the same way that they come by the general name and idea of man, they easily advance to more general names and notions. For observing that several things that differ from their idea of man, and cannot therefore be comprehended under that name, have yet certain qualities wherein they agree with man, by retaining only those qualities, and uniting them into one idea, they have again another and a more general idea; to which having given a name, they make a term of a more comprehensive extension: which new idea is made, not by any new addition but only, as before, by leaving out the shape and some other properties signified by the name *man*, and retaining only a body, with life, sense, and spontaneous motion, comprehended under the name *animal*.

Of Locke's account Jonathan Bennett (1971: 22–3) has remarked that the

theory of abstractness is required by Locke's view that the meaning of a word is determined by the idea (or class of similar ideas) associated with it . . . If I say of something 'That is an animal', giving this its ordinary meaning, I say nothing about what sort of animal it is; and so the 'idea of animal' in my mind must not be the idea of a vertebrate animal, nor may it be the idea of an invertebrate animal, since an idea of either of these kinds would endow my utterance with a stronger meaning than is ordinarily carried by 'That is an animal'.

And Bennett concludes,

if meanings are determined by ideas, then the fact that meanings can be more or less informative or specific implies that ideas must be able to be more or less saturated with detail. To mean what is ordinarily meant by

[3] For an unfortunate conflation of reference-generality and sense-generality, see Zwicky and Sadock (1975). For criticism of Zwicky and Sadock (1975), see Atlas (1977*b*: 328–30), L. D. Roberts (1984, 1987), and ch. 2, sect. 2, and ch. 3, sect. 1.

'animal', I must make it 'stand for' an idea which has enough detail to count decisively as an idea of an animal but isn't detailed enough to count as an idea of an *F* animal, for any non-vacuous *F*.

Thus, Bennett notes, "an 'abstract' idea" is an "idea such as one might have in imagining something, which is in some way sketchy or undetailed".

Hume and Berkeley thought Locke's abstract Ideas were psychologically impossible; we just couldn't have them. Worse, if one takes Locke to reify ideas so that an idea of triangularity is itself triangular, an abstract idea of triangularity would be logically impossible, since it would be neither equilateral nor scalene, etc. But the case for Locke's abstract ideas is not as weak as Berkeley thought. As Bertrand Russell (1945: 661–2) pointed out in criticism of Hume, we can have undetailed or sketchy mental images. Locke's example of triangularity, a rather special mathematical case, was one of the most difficult examples that he could have chosen, as Berkeley's criticism made evident. Locke's description of the case is justly notorious (*Essay*, IV. vii. 9):

> Does it not require some pains and skill to form the general idea of a triangle? . . . For it must be neither oblique nor rectangular, neither equilateral, equicrural nor scalenon; but all and none of these at once. In effect, it is something imperfect, that cannot exist; an idea wherein some parts of several different and inconsistent ideas are put together.

Some contemporary interpreters have put outrageously uncharitable glosses on this passage, e.g. Bennett (1971) and Staniland (1972). The first gloss is that Locke is here conceding that the general idea of a triangle is inconsistent, being all of oblique, equilateral, etc.; the second is that Locke admits explicitly that the idea cannot exist.

To take the latter point first: in 'it is something imperfect, that cannot exist', the 'that' is not a relative pronoun anaphoric with 'it'; it is a demonstrative pronoun beginning the appositive clause that refers to the hypothetical object, the "general triangle", a non-particular, not to "it" the idea of the object. Locke is not stating absurdly that there is no idea; he is merely stating that no "general triangle" can exist.

The former point suggests that Locke took his general idea to be an idea of an oblique, equilateral, etc. triangle. This gloss is supported allegedly by the final clause of the passage, where

Locke speaks of the idea "wherein some parts of several different and inconsistent ideas are put together". This remark, Bennett (1971: 37) says, is "wrong or irrelevant". Locke's last remark certainly need not be interpreted in so uncharitable a fashion. The emphasis should be on Locke's use of the word 'some': a general idea of a triangle is an idea where SOME parts of several different and inconsistent combinations of ideas of oblique, equilateral, etc. triangles are put together consistently. Indeed, consistency requires that only some parts are put together, and the resultant idea is "imperfect".

What does Locke mean by 'imperfect'? It is striking that the commentators ignore the question entirely. When I first studied this passage I believed that Locke meant "incomplete". The general idea of a triangle is incomplete, in just the sense that, by contrast, the oblique-triangle idea is, in respect of the triangle's shape, a complete idea. Commentators have taken Locke's phrase 'all and none' to mean that the general idea has all of the shapes and none of the shapes. But they have no literary sense; the phrase is a trope by which Locke attempts to express the way in which the abstract idea of triangle is shape-non-specific. It is, therefore, "imperfect" in respect of shape, i.e. the complex idea of triangle omits all simple ideas of shapes.

After having entertained this interpretation, I examined the 1671 Draft A version of the *Essay*, section 7 (Nidditch 1980: 47–8), where Locke used the terms 'perfect' and 'imperfect':

1 He that frames an Idea out of a partial collection of those simple Ideas that belong to it and leaves out several, hath an IMPERFECT [*my emphasis*] knowledg. as a child that hath noe other Idea of a man but a white colour & his shape made up of clothes & parts togeather.

2 He that frames an Idea consisting of such a collection of simple Ideas as are in that & belong not to any other subject hath a DESTINCT [*my emphasis*] knowledg, soe he that unites all these simple Ideas: bright yellow, weighty which is a comparison of motion downwards compard with bulke, fusibility, ductility, solubility in aqua regia, hath a DESTINCT knowledge or Idea of gold.

3 He that frames an Idea that consists of a collection of all those simple Ideas which are in any thing hath a PERFECT [*my emphasis*] knowledg of that thing but of this I must forbeare an instance till I can finde one.

4 He that frames an Idea of such a collection of simple Ideas as are INCONSISTENT [*my emphasis*] & are noe where united togeather hath noe knowledg of any thing there by but his owne imaginations.

It is clear from the last point (4) that Locke is not ignorant of the notion of inconsistency. In the first draft of the *Essay* a perfect Idea is a complex Idea that is a combination of all of the simple ideas of an object; an imperfect Idea is a complex Idea that is a combination of less than all of the simple Ideas of the object. Ultimately Locke will distinguish abstract Ideas from complex Ideas, but the notions of perfect and imperfect abstract Ideas, where perfection was originally a property only of complex Ideas, will linger on. Complex Ideas are "framed" from simple Ideas by a kind of "addition" or "supplementation"; abstract, general Ideas are "framed" from complex Ideas by a kind of "subtraction" or "deletion". These Ideas, unlike those of 'nurse' and 'mama', are not necessarily (and not even necessarily similar to) mental pictures, even on Locke's own view (*Essay*, III. iii. 7).

But where, historically, does this terminology of 'perfect' and 'imperfect' come from? The answer is, from Scholastic logic. For example, Thomas Cardinal Cajetan's *De Nominum Analogica* (1498), chapter 4, no. 36, states of an analogous, i.e. polysemous, term, that "we must distinguish a two-fold mental concept of the analogous name—one perfect and the other imperfect—and we must say that to the analogous name and its analogates there corresponds one imperfect mental concept, and as many perfect concepts as there are analogates" (Ross 1981: 213). Locke has merely adopted the relevant and familiar Scholastic vocabulary.

More to the point against Locke than Bennett's naïvely unhistorical criticism is the criticism that there are terms for which no mental image determines a condition for the application of the term—Do you really have an image of an invertebrate?—and the question, Even if some image comes to mind, how does that image give you the meaning of the term with which it is psychologically associated?

Suppose you could conjure up an image of an invertebrate. What is it in that image that tells you that things like that lack a backbone? If your "mental eye" inspects your mental image, does it examine the image with "X-ray" vision, as if there is something to be detected inside a mental image? If one says something like this at all, one must say that one's image of the invertebrate is initially like an X-ray picture. But then we repeat the original question: What makes the image in your mind just like an X-ray image of an invertebrate? Does it come labelled as such, like the ones of your stomach with your name written on the corner? The image must be itself seen as your stomach image; interpretation is required. But then we might as well have started with an image labelled 'invertebrate-animal image', if appending such a label were thereby to mean that the image to which the label is appended is to be itself described by that label. So it now seems, after these Peircian and Wittgensteinian pirouettes, that the label, or what the image is seen as, is the aspect of imaging that determines the meaning of the associated General term. So, at best, it was misleading of Locke to hold that an invertebrate-animal image determines the meaning of 'invertebrate animal'. In this form Locke's theory seems viciously circular, since recognizing an invertebrate-animal image as an invertebrate-animal image seems to require an understanding of the meaning of the term 'invertebrate animal'. And that is what Locke was trying to explain.

I wish to extract from Locke the view that some images are sketchy or undetailed, and the view that, as Russell (1945: 661–2) noted, it is psychologically possible to have such images. What Locke's theory does suggest, *pace* Mackie (1976), is that what we have recognized as an indeterminacy of the image is precisely analogous to what we have understood as sense-generality of the word, each indeterminate or unspecified with respect to some class of salient or theoretically "entrenched" pictorial and grammatical predicates respectively (Goodman 1983: 94). These are, *pace* Fodor (1983*a*), analogous characteristics of representations, inside or outside our heads, for visual perception and for verbal comprehension.

The problem that Locke leaves us with is the problem of the possibility of pictorial generality. (Recall that generality is lack of specification with respect to some predicate or other, or, to speak in Carnap's "material mode", with respect to some property or feature, typically some grammatical property or semantic feature in the case of the sense-generality of an expression.) The problem is, how can a picture be unspecified for some property, e.g. how can a triangle picture be unspecified for shape? Locke simply does not make such a possibility seem coherent. Either my mental, general, triangle image is nothing like any triangle picture that I could draw, any triangle picture having a specific shape and no "general" triangle image having a specific shape, or the mental image is like a drawn picture, and there are, *pace* Locke, no "abstract", mental triangle images.

It seems to me that in the *Essay* (III. iii. 7) Locke leaves it open whether abstract Ideas are pictures in the mind. I do not believe that he is committed to such a view. This allows him to sidestep the Berkeleian objection to "general" images. On the other hand, doing so makes the nature of mental images even more peculiar and mysterious. Since I think their nature is peculiar, and unlike that of well-behaved pictures, diagrams, etc., I am inclined to be sympathetic to Locke's intuition. The claim I wish to defend is stronger than Locke's: I wish to support the claim that there exist general pictures. Let us leave the introspectibles to one side and examine some physical objects, viz., paintings and drawings. (See N. Goodman and C. Z. Elgin (1988: 83–92) and Boden (1988: 27–44).)

Plane figures drawn without converging perspective offer no depth cues. Described by the Swiss crystallographer L. S. Necker in 1832, the classic example is a figure we take to display a cube (Fig. 1). The back face and front face are drawn the same size;

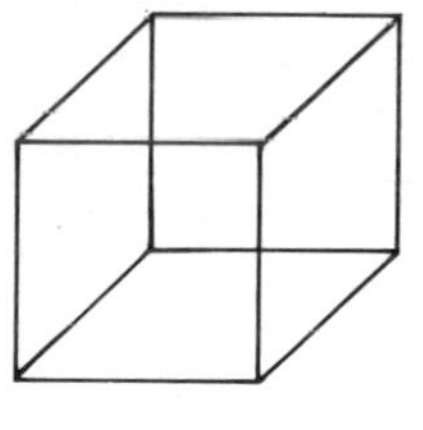

Fig. 1

thus no size difference—the smaller as the back, the larger as the front—serves to indicate which is the front, which the back. What is curious is that though we recognize the figure as a cube (and not as a truncated pyramid, which the figure would present if we always imposed converging perspective (Hochberg 1972: 56–7)—and I CAN see figure 1 as a figure of a truncated pyramid, the back face serving as the top, the pyramid being seen from above) what we see spontaneously reverses in depth. Alternately the figure is seen as a left-directed, downwards-projecting cube seen from above or as a right-directed, upwards-projecting cube seen from below.

Another figure of the same sort is a series of overlapping rings that can be seen as a tube whose near end is on the right, as a tube whose near end is on the left, or as series of rings in a plane (Fig. 2). Following Fred Attneave (1971) let us call such figures 'multi-stable'.

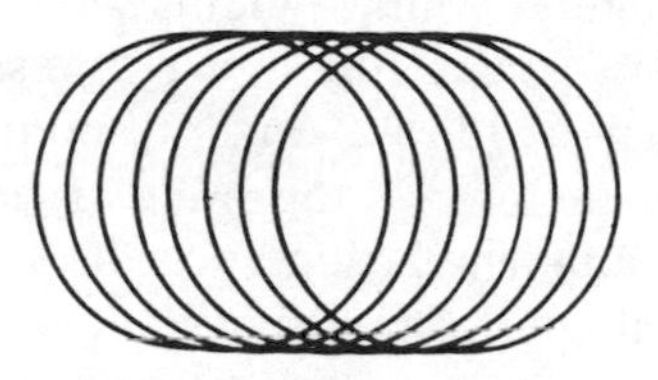

Fig. 2

Here is a story that Richard Gregory (1970, 1973, 1986) tells: Given an image on the retina, the brain tries to determine what object is being seen. Although the retinal image could be the result of any number of different projections onto a plane of any number of different three-dimensional objects, the brain either rejects or never considers odd or complex possibilities. The absence of perspective in the figure means that no unique solution to the brain's reduced problem can be computed, so it entertains in turn a small, finite number of probable solutions.

The remarkable depth reversals that attend these multi-stable figures dramatize what Sir Ernst Gombrich (1969) described as the Beholder's Share, the contribution made by the viewer in transforming the retinal image of a figure or picture into a presentation. The ring figure presents any of three "objects". Following Max Black (1972: 96–7) I shall say that what is "displayed" is analogous to a description's sense; what is "portrayed" by a picture, its actual subject, is analogous to a description's designation or reference. But unlike Black I shall say that what is "presented" is a content.

Psychologists have described multi-stable figures as depth-ambiguous. The ambiguity of 'the chicken that is ready to eat' is syntactic. The phrase manifests distinct grammatical analyses; it is the superficial outcome of encoding distinct syntactical roles for the Noun Phrase 'the chicken'. A multi-stable figure is not a figure that manifests distinct perspectival analyses; it is not the outcome of encoding distinct depth cues. Rather, it is constructed without conventional perspectival cues; the construction is depth-perspective free. So the figure is neutral with respect to depth perspective. This absence from the construction of the figure of perspectival cues for depth creates the multi-stability, the spontaneous reversals of depth in what is seen.

The ambiguity of a description differs from the spontaneous multi-stability of a figure. Unlike what is seen in the Necker figure as the mind tries to see what the figure "presents", the different senses of an ambiguous phrase do not spontaneously exchange places in the understanding as the mind tries to grasp what the phrase means. An ambiguous sentence does not spontaneously, freely, or periodically perform semantic flips. Nor does the grasp of one sense exclude the simultaneous grasp of another sense. These observations emphasize important differences between the mental processing of an ambiguous phrase and the processing of Figures 1 and 2.[4]

If my analysis of the depth non-specification of the Necker-cube drawing is correct, there should be a visual analogue to the Ross–Chomsky–Lakoff ambiguity test that I discuss in Chapter 2, Section 2. One ought to be able to see a conjoined, reduced, double Necker-cube drawing so that simultaneously one cube is seen in one perspective and the other cube in the other perspective. When I made this prediction in Jerry Sadock's presence, he promptly drew Figure 3. If the reader will fill in the two missing edges of that figure, he will discover that without visual gymnastics he will see two intersecting Necker-cubes in their

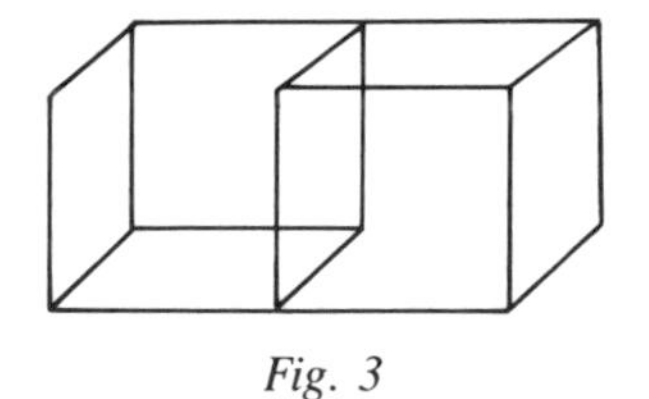

Fig. 3

[4] In fact, the claim that I have just made in the text needs to be qualified. I have discovered that by looking at a single Necker-cube drawing on a page along a plane making an intersection of a very small angle with the page (by holding the page nearly flat at eye-level and looking along it) instead of along a plane perpendicular to the page, in the usual manner, I can see the top of the cube-drawing in one depth perspective and the bottom of the drawing in the other perspective. So I must qualify my remark in the text by saying that *normally* the perception of one perspective excludes the perception of the other. (I have not encountered discussions of this phenomenon in the psychological literature. I would welcome references from a reader or research from interested psychologists.)

respective, different perspectives. The Necker-cube fails a spatial "reduced conjunction" test for depth ambiguity.

Lightfoot's (1982: 45) mistakes were in his claims (i) that the Necker-cube drawing is ambiguous, (ii) that a principle determining that certain interpretations are "usual" or "preferred", rather than "always" or "required", is adequate to explain the alleged complete absence of "unusual" interpretations, and (iii) that non-parallel interpretations of paired Necker-cubes require special visual gymnastics.

There are more subtle and complex kinds of multi-stability. If the camera does not lie, why are photographs often such bad likenesses of their subjects? What has the photograph failed to display? One immediate answer, of course, is movement. Sir Ernst Gombrich (1972: 17) writes:

> Clearly the artist or even the photographer could never overcome the torpor of the arrested effigy if it were not for that characteristic of perception which I described as "the beholder's share" in *Art and Illusion*. We tend to project life and expression into the arrested image and supplement from our own experience what is not actually present. Thus the portraitist who wants to compensate for the absence of movement must first of all mobilize our projection. . . . The immobile face must appear as a nodal point of several possible expressive movements.

How does an artist, be he painter or photographer, achieve by brush or camera a figure that will be a "nodal" point about which an image suggests movement? (Or to use some Gricean (1975) terminology, about which an image "implicates" movement?) Not, I believe, by creating an ambiguous figure.

One paradigm example of ambiguity is Joseph Jastrow's famous Rabbit-Duck figure (Fig. 4, after Attneave 1971: 94). Another is

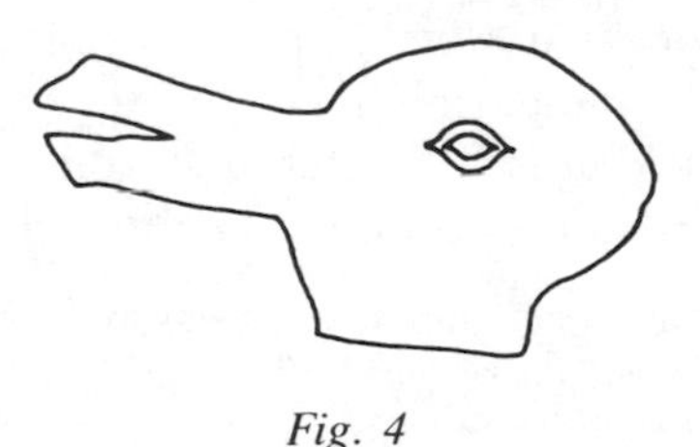

Fig. 4

W. E. Hill's Wife-Mother-in-law figure (Fig. 5, after Attneave 1971: 94). Like the alternative readings of an ambiguous sentence

Fig. 5

that are often unnoticed by listeners in conversation, the alternative displays of an ambiguous figure are often unnoticed by viewers. In both cases prompting is sometimes necessary before the alternative reading or display is perceived or comprehended. Then, once the several displays are recognized, like the ambiguous sentence whose different senses can be grasped simultaneously, the different "objects" displayed by an ambiguous figure can be seen simultaneously. For example, if I focus on the ' + ' in Figure 5 I can see Hill's figure simultaneously as the wife and as the mother-in-law. This is not typical of spontaneously multi-stable figures. It is the "split" aspects of multi-stable figures that make them SEEM properly describable as ambiguous, but, ironically, it is precisely this character that marks them psychologically as not ambiguous.

Can ambiguity serve the purposes that Gombrich describes? Gombrich (1972: 21) himself suggests not, when he writes, "The best safeguard against the 'unnatural look' or the frozen mask has always been found in the suppression rather than the employment of any contradictions that might impede our projection. This is the trick to which Reynolds referred in his famous analysis of Gainsborough's deliberately sketchy portrait style". Sir Joshua Reynolds wrote of Gainsborough (Gombrich 1969: 199–200):

> This chaos, this uncouth and shapeless appearance, by a kind of magic, at a certain distance assumes form, and all the parts seem to drop into their proper places . . . Gainsborough himself considered this peculiarity in his manner, and the power it possesses of exciting surprise, as a beauty in his works . . . I have often imagined that this unfinished manner contributed even to that striking resemblance for which his portraits are so

remarkable. Though this opinion may be considered as fanciful, yet I think a plausible reason may be given, why such a mode of painting should have such an effect. It is presupposed that in this undetermined manner there is the general effect; enough to remind the spectator of the original; the imagination supplies the rest, and perhaps more satisfactorily to himself, if not more exactly, than the artist, with all his care, could possibly have done.

When Reynolds said of Gainsborough that his "unfinished" and "undetermined" manner "contributed to that striking resemblance for which his portraits are so remarkable", he was standing at the edge of a truth about the way we faithfully, naturalistically, represent the world, whether on canvas or in words. Exactness is not the same as, nor does it entail, fidelity. (Conversely, fidelity does not entail exactness.) True-to-life portraits are "undetermined" and "nodal". They are like multi-stable figures, which are "undetermined" and "nodal" rather than ambiguous. Furthermore, those figures are "nodal" only if they are "undetermined". The fundamental feature is indeterminacy, i.e. pictorial generality. The artist's portraits and the psychologist's Necker drawings are respectively figures unspecified for "look" and figures unspecified for depth.

The same representational property of generality, i.e. representational "sketchiness", exists in both our naturalistic descriptions and our naturalistic depictions of the world. Naturalistically to render a subject rightly, we must talk and paint in a similar way. Portraits lack a particular "look"; expressions lack a particular element in their meaning. Gainsborough's portraits of his nephew Dupont and of Mrs Sarah Siddons are outside-the-head instances of Ideas abstracted from simple Ideas of Dupont and of Mrs Siddons, pictorial "abstract Ideas" corresponding to the Quinean General terms 'Dupontizes' and 'Siddonsizes'. The non-specification of the portrait figure for "look" both permits and delimits a range of seeings-as that constitute the range of "expressive movements" associated by Gombrich with the portrait. The possession of this range is what is implied by Gombrich's saying that the portrait figure is itself "nodal".

I have considered the Necker-cube drawing and the multiple-ring drawing as examples of pictorial depth-non-specification. I have suggested that the analogous fidelity of naturalistic portraits depends on a property of "look"-non-specification. In the cases of the Necker-cube and ring drawings, a general symbol severally "presents" distinct "objects" but is not ambiguous among them. To reinforce the lesson from the psychology of art, I shall now give an example of a figure that is pictorially even more general than the cube and ring drawings, yet is recognized by the viewer as a specific presentation.

First investigated by Italian psychologist Gaetano Kanizsa, the gappy, undetermined, "sketchy" figure "presents" a determinate but totally implicit perceptual object, a white triangle with illusory edges and brightness contrasts in front of a triangle with black edges (Fig. 6a, b, c, d; after Gregory 1973: 89). The white triangle is the result of the Beholder's Share; its visual presence is explained by Richard Gregory (1973) as the consequence of an interpretative strategy. The strategy accounts for the "gaps" in the black "triangle" and in the black "discs". They are apparently occluded by an overlapping, white "triangle". Thus the figure is recognized to "present" a white triangle in front of a black triangle and adjacent black discs. The sensation is a "predictive hypothesis" (Gregory 1986).

We do endow sense-general sentences and display-general figures with specific contents. Contrary to Mackie's (1976) and Fodor's (1983*a*) view, the verbal and visual systems in our heads do not talk utterly private languages. The representational systems we use in the visual mode are similar in at least some semantic respects to the systems we use in the verbal mode: symbols are semantically unspecified with respect to grammatical or pictorial properties, and apparently we possess a procedure for computing feature-specific outputs from a feature-non-specific input and collateral information. Neither art nor language renders the world rightly by copying it. If one or the other did copy its subject the rendering would not be right, or it would not be a rendering. A photocopy of a page, which may itself have been a photocopy of a page, does not describe or portray it; it duplicates it. A symbol system, verbal or pictorial, if it is naturalistically to render a world rightly (Goodman 1978: 127–40), must contain representations that are unambiguous but "undetailed" (as Bennett (1971) said of Locke), "sketchy" (as Bennett said of Locke, and Gombrich of Gainsborough), "unfinished" or "undetermined" (as Reynolds said of Gainsborough). They must have "nodal" meanings that are "dominant" (Gombrich) but not specific: that permit some concisions of sense and forbid others.

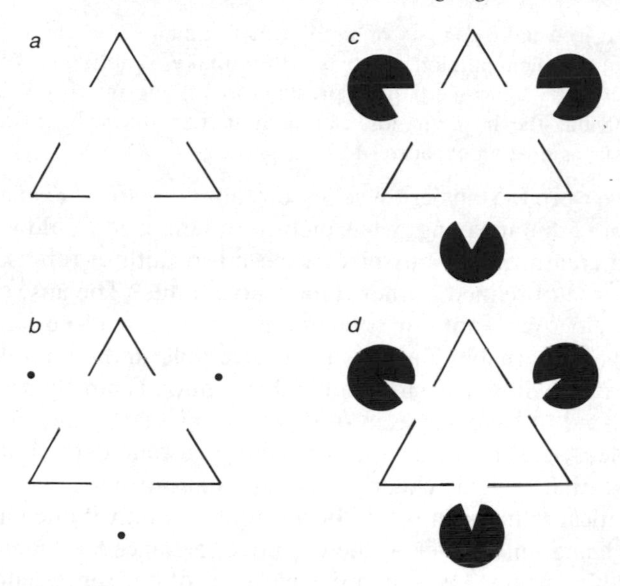

Fig. 6

Notes:
[a] A triangle with gaps.
[b] A faint illusory triangle may be visible, its corners touching the dots and covering the gaps.
[c] The illusory triangle is stronger. A photographic negative of this figure gives a darker-than-dark illusory figure.
[d] Smaller angles to the sectors produce a concave figure.

By emphasizing pictorial non-specification with respect to, e.g., depth, and "look", I have tried to evoke some intuitions about generality and its character. I have tried to illustrate the psychological difference between sense-generality and ambiguity. The Necker-cube drawing is general with respect to depth, i.e. it is depth-unspecified. The duck-rabbit drawing is ambiguous. The Kanizsa figure presents a triangle without presenting any of its sides; in some sense that vindicates Locke's intuition concerning the "all and none" character of abstract Ideas. And Gainsborough succeeds in life-likeness because he does not copy his subject; he sketches him. As Nelson Goodman (1978: 14) once wrote:

> The making of . . . [a] world . . . usually involves some extensive weeding out and filling in—actual excision of some old and supply of some new material. Our capacity for overlooking is virtually unlimited, and what we

do take in usually consists of significant fragments and clues that need massive supplementation. Artists often make skillful use of this: a lithograph by Giacometti fully presents a walking man by sketches of nothing but the head, hands, and feet in just the right postures and positions against an expanse of blank paper.

If we were to transfer these lessons about pictorial display to the case of verbal meaning, what picture of language would we have? Could creatures like ourselves succeed in faithful representation by the use of refined, rather than coarse, units? The answer seems to be 'No'. We want our conceptual schemes, and our languages, to have a learnable base, to have utterable and comprehensible statements whose transmission and reception is mostly immune to error, and to have a system of descriptions that is 'interrogatively complete' (i.e. for every question there is some correct answer. I believe that truth is cheap; it is explanation that is dear.) High semantical refinement would be without cost only if one had world enough and time as well as the cognitive tolerance for "interrogative incompleteness". The semantic character of our representations is a tell-tale mark of ourselves. We want an answer to every question, and a statement for every worldly occasion. If God's aim for our language were goodness-of-fit, answer-to-question and statement-to-fact, He would have had the sense to make our words sense-general. But we seem to have managed it on our own.[5]

[5] My indebtedness to Jacob Bronowski (1978), Nelson Goodman (1978), and to Friedrich Waismann (1945–6) is evident. I am also indebted to Sir Isaiah Berlin for an instructive conversation about Waismann.

2

Ambiguity and the Generality of Sense

0. The Fundamental Theses

In *Seven Types of Ambiguity* Empson (1930) discusses a Johnson poem, 'The Vanity of Human Wishes', used for illustrative purposes like mine by Jan Kooij (1971):

> *What murdered Wentworth, and what exiled Hyde,*
> *By kings protected, and to kings allied?*
> *What but their wish indulged in courts to shine,*
> *And power too great to keep, or to resign?*

The reader will observe that line 3 of this quatrain has an "ambiguity" that we mark by the following syntactic differences:

> [*their wish to shine*] [*indulged in courts*]
> [*their wish* [*to shine in courts*]] [*indulged*]

A further syntactical "ambiguity" concerns the readings of the elliptical 'their wish to shine indulged', where one may expand the phrases to 'their wish to shine indulged by themselves' or to 'their wish to shine indulged by others'. The elliptical phrase is syntactically neutral between these expansions.

Another type of ambiguity is lexical ambiguity. The verb 'indulge' can describe a dispositional or occurrent state, and so be classed as having the grammatical feature [*GENERAL*] or [*TEMPORAL*] (Kooij 1971: 122). There is also a lexical ambiguity in 'allied', viz. 'connected by marriage' and 'connected by treaty'. Both these cases are instances of polysemy, i.e. of different but related senses for one word. This is to be contrasted with homonymy, e.g. in the linguistic form *port*, where the same form realizes two words: '$port_1$' meaning "harbour", and '$port_2$' meaning "fortified wine" (Lyons 1977: 550). Ambiguities arising from word-sense or syntactic structure are "inherent" to the sentence. In understanding the poetry a reader selects one or more "readings" from those his knowledge of the language can provide him that seem to him appropriate in the poem.

Line 4 presents a rather different problem. 'Power too great to keep' can be understood (in specialized ways) as: (a) "power too great for Wentworth and Hyde to keep (but not too great for others to keep)"; (b) "power too great for anyone to keep". Likewise 'power too great to resign' can be understood as (a′) "power too great for Wentworth and Hyde to give up (but not too great for others to give up) willingly"; (b′) "power too great for anyone to give up willingly"; (c′) "power too great for Wentworth and Hyde to give up without danger from having exercised it"; (d′) "power too great for anyone to give up without danger from having exercised it". These are contextually relevant "specializations" of the generalized, literal sense of the words, distinct and relevant fillings-in, so to speak, of the sense of the words, but not themselves distinct senses of those very words. One is reading meaning into the words, not reading meaning out of them. In various contexts the word-tokens of *the girl with the flowers* can literally and felicitously be used to "present" the more specific contents "the girl wearing flowers", "the girl selling flowers", "the girl carrying flowers in her hand", "the girl strewing flowers", etc. Each of the relations between the girl and the flowers is subsumable under the [*WITH*] relation, but 'with' is not four or more ways ambiguous in sense (Kooij 1971: 110; Weydt 1973: 573; Brugman 1981). One should distinguish this phenomenon of *generality of sense* from lexical and syntactic ambiguity.

Finally, of course, 'the girl with the flowers' on different occasions of utterance will be used to refer to different girls. This constitutes for some a referential "ambiguity", along with the same sort of "ambiguity" for proper names.

After this brief survey of "ambiguities", let us consider an example of syntactical ambiguity in more detail, e.g. 'He hit the man with the fedora'. The prepositional phrase 'with the fedora' is either adjectival (an adnominal complement) in construction with the noun phrase 'the man', or adverbial (an adverbial complement) in construction with the verb phrase 'hit', specifically an adverbial of Instrument (Dillon 1977: 123). Parallel sentences are (Kooij 1971: 68) 'He cut the meat on the table' and 'They decorated the girl with the flowers'. Modestly transformed versions of these sentences will univocally express one of the meanings (Dillon 1977: 86). For example, the following have only the adverbial meanings: 'On the table he cut the meat', 'With the flowers they decorated the girl'. And these have only the adjectival meanings: 'The meat on the table was cut by him', 'The girl with the flowers was decorated by them'. Furthermore, similar prepositional phrases will occur in similar sentence frames but will not be ambiguous; they will express only the adverbial or only the adjectival meanings. For example, the following are univocal, the prepositional phrases having only the adverbial reading: 'He cut the meat quickly on a table', 'They decorated her lavishly with the flowers'. And the following are also univocal, the prepositional phrases having only the adjectival readings: 'The meat on the table is not fresh', 'The girl with the flowers is my niece'. After all, (i) if a sentence is syntactically ambiguous, the distinct structures that can underlie the same string of word-forms are structures that can uniquely underlie other and, often, superficially similar, strings of word-forms. It is also to be expected that (ii) when one string of word-forms can have, according to the grammatical rules of the language, one or another of two meanings, there will be distinct paraphrases of the string in the language, and sometimes, as in this case, quite close paraphrases, that univocally possess, according to the grammatical rules of the language, the distinct meanings. In such cases we call the rule-generated meanings 'readings', and I will abbreviate this relationship between the sentence-form and the meanings by saying 'the sentence form "displays" a reading'. In other words, a meaning that, according to the grammar of the language, is "displayed" by a sentence-form and that matches the literal meaning of some sentence-form in the language is dubbed 'a reading'. The assumption is that anything that can be meant literally by a sentence-form can be said literally in some sentence-form.

As I am using the expression 'meaning' a meaning is not a proposition, i.e. a reference-content, in the old-fashioned G. E. Moore-sense. So the meaning "displayed" in 'I am tall' is the same whether the sentence is uttered by John Walter or by me, since "display" depends solely on the grammar of the language. The literal meaning of 'I am tall' is neither "Jay Atlas is tall" nor "John Walter is tall". Assertoric utterances of the sentence 'I am tall' by me, i.e. my statements, "present" what is literally meant (i.e. "displayed") by the paraphrase 'Jay Atlas is tall', but those utterances do not "display" that Jay Atlas is tall. The meaning displayed by the sentence 'I am tall' is, roughly, "[*FIRST PERSON*] is tall". What is "portrayed" by my assertion of 'I am tall' is the state of affairs that Jay Atlas is tall.

In David Kaplan's (1979) terminology, what is displayed is a character of the sentence-type; what is presented is a content of a sentence-token. In Frege's terminology, though not his usage, what is displayed is a *Sinn*; what is presented is a *Gedanke*. In Stalnaker's (1972) jargon, what is displayed is a meaning; what is presented is a proposition. I prefer to put the emphasis on the presenting versus displaying by a sentence-token rather than to bifurcate "meanings" into two kinds, sentence-meanings and utterance-meanings. For me what is important is the relationship between meaningful sentences, those theoretical entities described by the grammar of a language, and the contents they can represent (to use an umbrella word): the ways they represent the contents they do. For others what has been important is the thing representing and the thing represented, i.e. the ontological character of the relata, e.g. meanings, propositions, states of affairs, facts. This metaphysical preoccupation has led to errors in philosophical speculation about the nature of language and to false doctrines in logical theory. I prefer to focus on the *modes of representation*: *display, presentation, and portrayal*. Some sentences, and for that matter pictures, are "ambiguous", i.e. (i) possessing distinct syntactical analyses, (ii) lexically homonymous, or (iii) lexically polysemous. By contrast, other sentences are (iv) sense-general or (v) multiply-referential in contexts of use. Still others are (vi) used indirectly to "present", through pragmatic inference (e.g. Grice 1975), contents that they do not "display". In each case we are concerned with a different relationship between a literally meaningful sentence, a context of utterance, and a content. The content is represented by an utterance, a token of the sentence in a context, understood according to the rules of the grammar and to the conventions of interpretation shared by language-users of the language of the sentence. The important question for my theoretical enterprise is not what, or what sort of thing, a sentence-meaning is, but *when* and *how* a meaningful sentence means; in this attitude I follow Nelson Goodman's (1978) lead.

My question has been: In interpreting utterances are we *selecting* from the linguistically given readings of a syntactically or lexically *ambiguous* sentence, or are we *constructing* from a meaningful but radically *sense-general* sentence a contextually determined interpretation of an utterance, an interpretation whose content is far more specific than the literal meaning of the sentence?

My answer has been: In interpreting utterances we are, more often than philosophers and even linguists have recognized, doing the latter. This essay, like my earlier work (Atlas 1974, etc.) defends this answer and examines the consequences of this answer for some of the central problems in the philosophy of language and of logic in twentieth-century analytical philosophy.

I have used the examples of indexical sentences, e.g. 'I am tall', to motivate a familiar distinction between (sense) ambiguity and multiple referentiality in contexts. 'I' is not (sense) ambiguous, though its reference shifts from user to user. Likewise, the sentence 'I am tall' is not ambiguous, though the propositions its utterance may be used to express differ from context to context. But this is not the phenomenon that interests me, though I hope that the reader will find it familiar and so heuristically useful. I do not think that the reader will be surprised that 'I' is not ambiguous in sense, even though its reference varies. The phenomenon that interests me is linguistically even more basic: the univocal 'I' is *sense-unspecified for gender*. Gender is irrelevant to the meaning of 'I', though relevant to the meanings of 'he' and 'she'. Similarly, gender is irrelevant to the meanings of 'cousin' or 'neighbour'. Because gender is irrelevant to the senses of these words, sex is irrelevant to the correct, literal applications of such words to objects. 'I' is equally at home in my mouth as in Kay Barbara Warren's mouth, since the genderless 'I' is semantically indifferent to the sex of whomsoever mouths it. And it is this phenomenon of semantic indifference, not the familiar phenomenon of referential indexicality, that interests me.

I have already remarked that the existence of various interpretations (contents) of 'with', appropriate to different contexts in which 'the girl with the flowers' is uttered, does not entail that 'with' is ambiguous in sense among "the girl wearing flowers", "the girl selling flowers", etc. These contextual specifications are not listed in my dictionary, the one on my desk or the one in my head. I do not search an antecedently given list for a sense that fits the context of utterance and *select* it as the interpretation that 'the girl with the flowers' presents in that context. Rather, it seems that knowing the meaning of 'with', knowing what 'the girl with the flowers' displays, I understand the fitting interpretation of 'the girl with the flowers' in the context to be, e.g., "the girl wearing the flowers". Rather than select a specific sense, I *construct* a specific interpretation.

Perhaps I may reinforce the point with different examples from Ruth Kempson (1977: 132–5) and from Kent Bach (1982: 593). Kempson's example is a negation. The sentence 'It wasn't a woman that came to the door' may be used to state a proposition that would be true if a girl came to the door, or a proposition that would be true if a man came to the door, but it is not ambiguous in sense merely because it may express the distinct propositions "It was a non-adult, human female that came to the door" and "It was an adult, human non-female that came to the door". The sentence is sense unspecified, not ambiguous, between these interpretations; the meaning of the sentence is *neutral* between them. The *meaning* of the sentence is *neither one nor the other*. A similar observation is made by Kent Bach (1982: 593):

> Jack tells Jill, 'I love you too'. He could mean any one of several things, (a) that just as Jill loves him, so he loves her, (b) that he loves Jill and someone else too, (c) that like someone else, he too loves Jill, or (d) that he has love as well as some other feeling for Jill. Whatever he may mean, he would be speaking literally and yet, I claim, the meaning of the sentence he is using, though univocal, does not fully determine what he means in using it. The point is not merely that what he means is a matter of his communicative intention, but that the sentence is semantically nonspecific. It* does not have a definite truth condition, even with Jack and Jill fixed as the values of 'I' and 'you'. A condition necessary for its* truth is that Jack love Jill, but one of four other conditions is necessary as well, depending on which of the (a)–(d) is meant. However the sentence is not thereby ambiguous, but merely semantically nonspecific.

I have asterisked the pronouns in order to emphasize the difficulty of saying without awkwardness what both Bach and Atlas (1975*b* etc.) want to say: *a sense-general sentence may, in context, "present" a proposition with definite truth-conditions but does not "display" a proposition*. Its sense is not to be identified with a proposition, and if a Russell–Tarski-style logical form has a definite truth-condition, the meaning of the sense-general sentence cannot be represented by the logical form. Thus I shall avoid unqualified talk of 'its truth', i.e. the *sentence's* truth.

These are issues to which I return at length later in this essay. Here the point is to reinforce the lesson of Bach's example. When

Jack tells Jill 'I love you too', his *sentence* is not four-ways ambiguous. Jill does not check a mental list of possible readings and *select* one (or more) that fits her context of utterance. In order to understand what Jack said, she *constructs* from contextual information and a sense-general sentence a specific interpretation, e.g. Bach's (a), (b), (c), or (d). The sentence is not ambiguous; it is univocal, and its meaning is perfectly *definite*! It is merely that this definite meaning of the sentence, the product of word-meanings and the syntactical rules of their combination, is *neutral* among interpretations (a), (b), (c), and (d). The meaning of the sentence is identical to none of them.[1]

A thesis that I defend in this book is:

> *The sense-generality of a sentence radically underdetermines (independently of indexicality) the truth-conditional contents of its utterances.*

Or, to put it another way, the sense of a sense-general sentence is not a proposition (the bearer of determinate truth or falsity). A correlative thesis is:

> *Sense-general sentences do not have Russellian logical forms.*

My third thesis is:

> *In interpreting the utterances of sense-general sentences, we are not selecting from the readings of an ambiguous sentence; we are* CONSTRUCTING *from a definite but general sense and from collateral information a specific content.*

In the course of this book I shall expound and defend these Fundamental Theses.

1. Criteria for Ambiguity and Generality of Sense

Sides of rivers and financial institutions are denoted by the word form *bank*, but is the word-form a form of one word with two meanings, or is it a form instantiating two words? If there are two etymologies, linguists are tempted to opt for two words. If there are two grammatical classes and unrelated meanings, they also usually opt for two words. But Quine (1960: 130, and again in his

[1] It is also not identical to the logical disjunction of them, as I shall show in Chapter 2, Section 3.

new introduction to Quine (1980)) chooses to identify the word with the word-form, determined by sound or shape: "For our purposes, matters may most easily be kept straight by calling words identical that sound alike (or look alike, if writing is in question)". This convention is consistent with the Tarskian tradition in formal logic in separating syntactic objects from their (model-theoretic) interpretations.

When a singular term, e.g. the proper name 'Paul', severally designates thousands of Pauls, Quine describes the name as ambiguous. It is noteworthy that he uses 'ambiguous' to describe multiple reference, not multiple sense.[2]

In the case of General terms, e.g. 'light', Quine notices that one may truly state 'Dark feathers are light' and falsely state 'Dark feathers are light' (or, 'Dark feathers are light, and dark feathers are not light'). The law of non-contradiction is saved by the ambiguity of 'light', which is manifested by the term's being simultaneously true and false of the same things. This is a very narrowly focused criterion of ambiguity. Quine (1960: 130) observes:

> in the case of an admittedly general term, how are we to say how much of the term's multiple applicability is ambiguity and how much generality? Take 'hard' said of chairs and questions. As remarked, ambiguity may be manifested in that the term is at once true and false of the same things. This seemed to work for 'light', but it is useless for 'hard'. For can we claim that 'hard' as applied to chairs ever is denied of hard questions, or vice versa? If not, why not say that chairs and questions, however unlike, are hard in a single inclusive sense of the word? There is an air of syllepsis about 'the chair and questions were hard', but is it not due merely to the dissimilarity of chairs and questions? Are we not in effect calling 'hard' ambiguous, if at all, just because it is true of some very unlike things?

For Quine (1960: 131), a term's being clearly true or clearly false, from utterance to utterance, of one and the same thing,

> if not a necessary condition of ambiguity of a term, is at any rate the nearest we have come to a clear condition of it. We have taken account of ambiguity only insofar as it figures as a contributing cause of variation in the truth value of a sentence under variation of the circumstances of utterance.

Since criteria for ambiguity were discussed at length by Aristotle in the *Topics*, not to mention by the Stoics, the Scholastics, and

[2] See Cohen (1980) and Strawson (1980).

philosophers of the modern period, Quine's seems a modest effort. Characteristically Quine's criterion is extensional, defined as it is by variations in truth-value. Of course Quine (1960: 132) is perfectly aware of the modesty of his criterion: "The shifting of the reference of 'the door' and of the truth value of 'the door is open' with circumstances of utterance are accounted normal to the meanings of the words concerned, whereas ambiguity is supposed to consist in indecisiveness between meanings." But, as he goes on to write, "Our reflections in Chapter Two ["Translation and Meaning", of *Word and Object*] encourage us little in distinctions of this kind . . .". Since the reflections in question are founded on methodology characteristic of American social science in the 1920s and 1930s, e.g. J. B. Watson (1924), L. Bloomfield (1933), and Skinner's pigeon-training at Harvard, Quine's philosophical inability to countenance the distinction between truth-conditions and meaning is a linguistic *reductio ad absurdum* of his philosophical behaviourism. A difference in truth-value is obviously not a necessary condition for a difference in meaning, and one is not limited to shifts in the reference of a singular term to demonstrate it. There are sentences of English that are ambiguous, but whose distinct readings are none the less mutually entailing, and so the readings have the "same" truth-conditions; to take an example from J. L. Morgan, 'Someone is renting a house'.[3] A person unable to accept the distinction between truth-conditions and meaning is typically, like Morton White, Quine, or Donald Davidson, a philosopher with a particular epistemological programme. But Quine cannot forswear talk of senses either, e.g. in Quine (1960: 132): "A particularly prominent species of the syncategorematic use of adjectives is that in which an adjective that admits of comparison, e.g. 'big', is used with a substantive in the fashion 'FG' to express the *sense* 'G that is more F than the average G'; thus 'big butterfly'", or again, "One extends the notion of ambiguity beyond terms to apply to particles—notably 'or', with its proverbial inclusive and exclusive *senses*—and even to syntax" (Quine (1960: 134)). In the case of 'or', Quine's sense might be truth-conditions, but he does not say so. Whether

[3] This counter-example also refutes Cruse's (1986: 82 n. 12) same necessary condition for ambiguity. His criticisms of Lyons's (1977: 404) and Kempson's (1977: 128–9) arguments seem to me correct none the less. He has good objections to bad criticisms of a false condition that unfortunately he accepts.

Quine's semantic ascent mounts higher than the foothills of reference is thus, surprisingly, a moot point. But Quine's univocalism is not an open question. He stoutly asserts it in the following remarks (Quine 1960: 131):

> There are philosophers who stoutly maintain that 'true' said of logical or mathematical laws and 'true' said of weather predictions or suspects' confessions are two usages of an ambiguous term 'true'. There are philosophers who stoutly maintain that 'exists' said of numbers, classes, and the like and 'exists' said of material objects are two usages of an ambiguous term 'exists'. What mainly baffles me is the stoutness of their maintenance. What can they possibly count as evidence? Why not view 'true' as unambiguous but very general, and recognize the difference between true logical laws and true confessions as a difference merely between logical laws and confessions? And correspondingly for existence?

Ryle (1949) offered a linguistic criterion for ambiguity that goes beyond the sufficient condition that Quine considers, viz. that if 'John is a nut' and 'John is not a nut' are both true, 'nut' is ambiguous. At least Ryle noticed, correctly, that 'Jay and Snoopy are lovable characters' cannot be understood to mean 'Jay is a real, lovable eccentric and Snoopy is a lovable fiction'. It either means 'Jay is a real, lovable eccentric and Snoopy is a real, lovable eccentric', or it means 'Jay is a lovable fiction and Snoopy is a lovable fiction'. Since on Ryle's view Snoopy and I are of different logical types, if the term 'lovable character' applies truly or falsely to both of us it does so with different meanings. But the sentence 'Jay and Snoopy are lovable characters' linguistically requires that 'lovable character' have the same meaning. Since, on Ryle's view, 'lovable character' cannot in the same sense be applied, truly or falsely, to me and to Snoopy, the sentence 'Jay and Snoopy are lovable characters' is logically unacceptable. Ryle begins from a correct linguistic observation: if *a and b are F* contains an ambiguous *F*, say meaning "F_1" and "F_2", *a and b are F* has two readings: (i) "a is F_1 and b is F_1"; (ii) "a is F_2 and b is F_2". The sentence cannot mean "a is F_1 and b is F_2"; and it cannot mean "a is F_2 and b is F_1".[4]

In the grip of Russell's Theory of Types, Ryle assumes that the

[4] There are obvious restrictions on substitutends for *F* in the surface-structure schema. For example, we shall not let the metavariable *F* take as a value the English phrase 'each other's keeper'. See the Appendix for further discussion of Ryle (1949).

same "linguistic form" predicated significantly of each must have different senses in the two predications. If so, he would expect this logical restriction on the available senses of a "linguistic form" in a significant predication to yield a linguistic anomaly in a sentence whose grammatical restrictions on its available readings force a "logically meaningless" predication on an individual hungering for significance. Our quarrel, if any, must be with Ryle's Russellian assumption not his linguistic intuition. Ryle has provided an answer to Quine's question 'What can they possibly count as evidence?' Quine has seemingly ignored it.

Quine's other question is, How much of a General term's applicability is ambiguity and how much "generality"? His answer, for 'true' and for 'exists', is: It's all "generality". (I am myself persuaded by Morton White's (1956) arguments that this is the best answer for 'exists': see the Appendix. For the moment I wash my hands of 'truth'.) 'Exists' in its multiple applicability denotes physical objects and abstract objects. Quine's question is whether classes of physical objects and classes of abstract objects are subclasses of *the* extension of 'exist', or whether they themselves constitute distinct extensions of 'exist' like different Pauls named by 'Paul'. Is the General term 'exist' "general" or "ambiguous", a term of heterogeneous extension or a term whose different extensions are homogeneous? Quine couches his discussion in terms of reference and extension. His notion of "ambiguity" is an extensional one, and so is his co-ordinate notion of "generality". Famously, Quine restricts himself to the theory of reference. This creates a terminological problem, since Quine admits that his notion of "ambiguity" is not "indecisiveness between meanings" (Quine 1960: 132). His "ambiguity" could perspicuously be called 'distinct referential specification'. His "generality" could perspicuously be called 'referential generality'. 'Exist' is referentially unspecified between physical and abstract objects, so that it has one extension that contains both.

Contrary to Quine's practice, it is scientifically better to preserve both sense-concepts and reference-concepts. A General (contrasted with Singular) term is general in sense (contrasted with ambiguous) when its sense is unspecified for a semantic feature [F], e.g. a gender feature that specifies the meaning contrast of 'mother' and 'father'. A General term is referentially general when the presence or absence of a property of its denotations does not affect the identity of its extension, e.g. the sex of mothers or fathers. As we have already seen, generality of sense is always relative to some semantically "basic" predicate of the lexicon, some semantic feature [F], and should be understood as '[F]-general' for some [F], though typically I leave the reference to an [F] implicit. Likewise, referential generality would be relativized to some property Φ and should be understood as 'Φ-general' for some Φ, though again I leave the reference to a Φ implicit. On White's (1956) and Quine's (1960) view, 'exist' is [ABSTRACT]-general in sense and abstractness-general in reference.

Lately the question of sense-generality versus ambiguity has been raised in linguistics. Green (1969: 81) has argued that statements like

(1) *Charlie refused Mort's request.*
(2) *Charlie refused Jerry's offer.*

suggest that 'refuse' is polysemous, having two readings 'refuse to grant' and 'refuse to accept'. Ruhl (1975) replied, in a Quinean spirit, that the difference in interpretation between (1) and (2) arises from the difference between requests and offers, not from an ambiguity in 'refuse'. Then Ruhl, like Gilbert Ryle (1949) and James McCawley (1968), invokes a Reduction Test, which permits grammatically acceptable deletions only of elements that have the same meaning. For example, 'light' has the same meaning in:

(3) *Charlie is light.*
(4) *The box is light.*

This permits the acceptable reduced form:

(5) *Charlie is as light as the box.*

If Green were correct that 'refuse' is ambiguous, reduced forms such as the following should not be grammatically acceptable:

(6) *Charlie refused Mort's request and Jerry's offer.*

Ruhl (1975: 201) points out that a statement like (6) is acceptable, and so it is doubtful that 'refuse' is polysemous. The difference in one's understanding of (1) and (2) depends on one's knowledge of the difference between requests and offers, not on a lexical ambiguity of 'refuse'.

Quine's term 'ambiguous' in Quine (1960) was just a neologism. If we generalize the notion of referential generality beyond terms, i.e. beyond *n*-place predicates ($n \geq 1$), to sentences, i.e. 0-place predicates, we encounter familiar views. If we associate with a statement truth-conditions realistically understood as states of affairs in which the statement would be true and states of affairs in which the statement would be false (and, if a "vague" statement, states of affairs in which it would be neither), the referential generality of a statement amounts to a lack of specification of the "kinds" of states of affairs that would verify or falisfy the statement. Whatever way the "kinds" are sorted, e.g. the mathematical kind contrasted with the spatio-temporal kind, the distinction between kinds bears not at all on the verification or falsification of a statement that is referentially unspecified with respect to those kinds. To take a trivial example, if *John kicked Brian* is true, it is true because John kicked Brian with his (John's) right foot, or because John kicked Brian with his left foot, or, more impressively, because John kicked Brian with both feet simultaneously, or . . . The distinctions just drawn are not linguistically relevant in English; 'kicked' leaves such matters open. So one can say that *John kicked Brian* is referentially unspecified, i.e. truth-conditionally neutral, between "instrumentally" distinguishable states of affairs in its truth-conditions.

This is a familiar, utterly unsurprising linguistic fact of meagre theoretical interest. The notable point here is that on Quine's way of putting the matter the semantical alternative would be that *John kicked Brian* is "ambiguous" (in Quine's extensional sense) between, say, left-foot kickings and right-foot kickings. This is a *reductio ad absurdum* of Quine's way of putting the matter. No word seems more out of place here than 'ambiguous'. So I shall abandon philosophical jargon and adopt linguistic terminology. I shall distinguish reference-non-specification from reference-specification with respect to kinds of extensions. I shall distinguish this distinction in the theory of reference, a distinction that Quine obviously defends, from a parallel distinction in the theory of meaning, which Quine obviously ignores. That distinction is the distinction between sense-generality and multiple-sense-specification, or more familiarly put, between sense-generality and (real, sense-)ambiguity. I shall take meaning seriously, which is to say, in Chomsky's sense of 'syntax', I shall take syntax seriously.

Many distinct states of affairs would make true my utterance *Charlie is in Wyoming*, but this multiplicity of verifications does not entail a multiplicity of meanings of the sentence 'Charlie is in

Wyoming' (see Chapter 2, Sections 2 and 3; Katz (1980: 27–8)). It is natural then to inquire, what does ambiguity consist in? If a multiplicity of verifications isn't sufficient for ambiguity, what is? A natural thought to have is that paraphrases will provide evidence of ambiguity. After all, in the examples of *bank*s, the word-forms are forms of lexemes meaning "money depository" and "side of a river". This is an easy case. Suppose we take 'brother-in-law' instead (Lyons 1977: 404), and paraphrase it by 'wife's brother', 'husband's brother', 'sister's husband', etc. Does this mean that 'brother-in-law' is an ambiguous expression? I take it to be obvious that these are referentially based paraphrases, descriptions of the denotata of the expressions from the kinship vocabulary of English. We are back with the multiplicity of verifications.

The word-form *bank* in English could be translated into French by 'banque' and by 'rive'. Would it be sufficient to show that an English expression were ambiguous if it were translatable into distinct lexemes in another language? I shall take Lyons's (1977: 404) word for it that 'brother-in-law' can be translated into four non-synonymous lexemes in Russian. So, even if such translation is suggestive, prima facie evidence for ambiguity, it is not conclusive. It will not distinguish ambiguity from what Lyons (1977: 404) calls 'generality of sense'.[5] Despite the different Russian lexemes, 'brother-in-law' is not four-ways ambiguous in English.

If simple paraphrase and translation are not conclusive tests for the presence of ambiguity, perhaps something a little more linguistically sophisticated will be. I think it not unfair to say that almost all recent philosophical claims for the ambiguity of sentences have rested on simple paraphrase and translation criteria. Sometimes the criterion has even been paraphrase into a regimented, formal language. The extensional character of model-theoretic interpretations of the well-formed formulae of these formal languages implies that the multiplicity of interpretations is at best a multiplicity of verifications (truth-conditions). I have already rejected the view that the multiplicity of truth-conditions implies a multiplicity of meanings. Now, in addition, we see that

[5] The first discussions by linguists that were accessible to philosophers were Zwicky and Sadock (1975) and Lyons (1977). Since the appearance of Lyons (1977), in particular, philosophers have had the opportunity to reflect on these matters. For the most part they have not done so.

even intuitively acceptable paraphrase in the same natural language or translation into another natural language is not conclusive evidence for ambiguity (even assuming a conventional translation manual).

Gilbert Ryle (1949) was more acute than his successors in this matter. His linguistic arguments were, in retrospect, an impressive exercise in method. He had explicitly appealed to the kind of criteria that linguists have since 1957 found compelling. But let us consider a syntactic rather than lexical example, which I take from John Lyons (1977: 405), whose lucid account I shall be expounding and criticizing here.

Conjoined forms should have the same distribution, and so grammatical role, in constructions in the language. If they do not, conjoining them will produce odd sentences, i.e. zeugma, e.g. 'We heard your voice and him slam the door'. Obviously the noun phrase 'your voice' and the string 'him slam the door' have different distributions in sentences of English and are not the same sort of grammatical "object" of the transitive verb. If we substitute 'her cry for help' for 'your voice', we produce the acceptable 'We heard her cry for help and him slam the door'. If we substitute 'her cry for help' for 'him slam the door', we produce the acceptable 'We heard your voice and her cry for help'. In each case the interpretation of the co-ordinate structure must be parallel, either two noun phrases or two object-pronoun–verb phrases.

This makes it clear that the utterance *We heard her cry for help* is structurally ambiguous. It might also be lexically ambiguous, if we distinguish different lexemes in the strings of 'We heard her cry for help', distinguishing 'cry' as a noun from 'cry' as a verb. In addition, if the verb 'cry' is polysemous, having the meanings "weep" and "shout", or if one '$\text{cry}_{[\textit{weep}]}$' is homonymous with another '$\text{cry}_{[\textit{shout}]}$', then *We heard her cry* will be lexically ambiguous.

The interesting question, and the one originally motivating Ryle (1949), Morton White (1956), and Quine (1960), has the form: Is 'cry' polysemous or is it sense-general? In particular, is the relation of 'cry' to 'weep' and 'shout' one of a sense-general verb to various specifications, like 'red' to 'scarlet' and 'crimson'? Or is 'cry' polysemous, with two senses?

We might try to shed light on this question of lexical ambiguity by adopting the criterion of reduction that assisted us in determining the structural ambiguity of 'We heard her cry for help'. *If 'cry' is sense-non-specific for those aspects of meaning that determine in different ways the more specific meanings of 'shout' and 'weep', it is neutral in sense between the two. This is the basic intuition concerning sense-generality*.

If 'cry' is two-ways ambiguous, on the other hand, the conjunction 'John cried and Brian cried' is in principle four-ways ambiguous. A reduced form requires sameness of sense for elements to have been deleted; so, if 'cry' is ambiguous, 'John cried and so did Brian' should possess only parallel readings. The reduced form in fact has no crossed interpretations "John wept and Brian shouted" or "John shouted and Brian wept". It simply does not have those meanings. 'John cried and so did Brian', and similarly 'John and Brian cried', can only literally mean "John wept and Brian wept" or "John shouted and Brian shouted". It is clear that if an expression is ambiguous the sameness-of-sense condition on deletion implies that the reduced form of the conjunction will have only parallel readings. (Parenthetically, one might wonder what the linguistic evidence for the general sameness-of-sense condition is. SOME of the evidence for the condition was, presumably, some pre-theoretical intuitions about the number of readings of reduced conjunctions containing assumed-to-be-ambiguous expressions.) But, if one is wondering whether an expression is ambiguous, evidence for the ambiguity will be the unavailability of crossed interpretations as possible literal meanings of the reduced sentence. If the evidence is regarded as theoretically conclusive, we can say:

(A) *The impossibility of a crossed, literal paraphrase for a conjunction-reduced sentence S entails the ambiguity of S.* (The distinct, parallel paraphrases express distinct senses.)

This criterion of ambiguity should be sharply distinguished from a superficially similar claim:

(*) The oddity of a conjunction-reduced sentence *S* entails the ambiguity of *S*.

Ryle (1949) said of 'She came home in a flood of tears and a sedan chair' that its infelicity implied that the expression 'in' was ambiguous. This example is not a case for criterion (A), which applies to acceptable sentences and demands an intuitive judgement of whether a sentence *S* can literally mean *P*. Criterion (*) applies to "odd" sentences and ASSUMES that the infelicity is ONLY explained by the ungrammaticality of a reduced sentence-form, whose apparent interpretation, even if the sentence-form is ungrammatical, must be an impossible crossed reading. Interpretations of a reduced, co-ordinate expression that are possible literal paraphrases of the expression are interpretations with identities of sense. Obviously the converse does not follow, that interpretations with identities of sense will be interpretations that are possible literal paraphrases of an expression, and hence acceptable renderings of the senses of the expression. So the contra-positive of the converse does not follow, that an interpretation that is "odd" or unacceptable is an interpretation with non-identities of sense. Thus, zeugma does not entail ambiguity.

The reliability of the co-ordination test for ambiguity, i.e. the adequacy of criterion (A), is sometimes thought, even by Lyons (1977: 407–9), to depend on acceptability judgements on sentences. Since these intuitive judgements are notoriously dependent on verbal and situational contexts in which tokens of the sentence occur, the criterion is no more reliable, it is usually argued, than these varying acceptability judgements. In general it is useful to keep the variability of acceptability judgements in mind, but as Lyons (1977: 406) has formulated (A) this variability is irrelevant. In (A) the conjunction-reduced sentence *S* is presumed to be acceptable.

It is also worth noting the difference between (A) and its converse (B):

(B) *The possibility of a crossed, literal paraphrase for a conjunction-reduced sentence S entails the non-ambiguity of S.* (The distinct, parallel paraphrases do not express distinct senses.)

Obviously (A) does not entail (B). Lyons (1977: 408) independently accepts (B) as well as (A).

Now I wish to discuss an application of criteria (A) and (B) to an example, an application that Lyons himself finds contentious. The question Lyons raises is whether 'like' is polysemous as opposed to sense-general in utterances like:

(7) *Charlie likes brunettes.*
(8) *Charlie likes daiquiris.*

If simple paraphrases were more decisive than they are, we would be pushed toward the polysemy view by paraphrases like:

(9) *Charlie has friendly or amorous feelings for brunettes.*
(10) *Charlie has positive feelings about drinking daiquiris.*

On the other hand the generality view might be recommended by the paraphrases:

(11) *Charlie has a pro-attitude toward brunettes.*
(12) *Charlie has a pro-attitude toward daiquiris.*

the difference being not in the character of the attitude, i.e. in the meaning of 'likes', but in the character of the objects of the attitude. The reduction test gives us some help when confronted with this choice. Consider utterances like:

(13) (*a*) *Charlie likes brunettes and daiquiris.*
(*b*) *Charlie likes brunettes and Jay daiquiris.*
(*c*) *Charlie likes brunettes more than daiquiris.*

Lyons (1977: 408) suggests that most speakers of English would find that these utterances are all prima facie "odd". Certainly there is something mildly unexpected in the conjunctions. The question for Lyons is whether the surprise is a linguistic one. Does the tendency to impose semantic parallelism in interpreting these utterances produce a linguistic anomaly, because parallel paraphrases are unavailable for these sentences in the case of polysemy? If there is anomaly, are polysemy and the semantic impossibility of 'likes$_{[daiquiris]}$ brunettes' and of 'likes$_{[brunettes]}$ daiquiris' the explanation of it? Anomaly and polysemy aside for the moment, is there any SEMANTIC impossibility here?

It is important to note, as Lyons (1977: 407) does, that the bizarreness of amorousness about daiquiris is not so obviously a semantic bizarreness as a psychological one, and the bizarreness of drinking a brunette is as much the bizarreness of a blender large enough and the character of one's diet. Even in the case of polysemy, parallel paraphrases are linguistically possible. Now, how much anomaly is there? Would MOST native speakers of English find the sentences anomalous? I tend to doubt it. I do not find linguistic anomaly here.

Lyons (1977: 408) suggests that the first and the second sentences can be appropriate in some contexts, and appropriate interpretations are not parallel. That is to say, an utterance of 'Charlie likes brunettes and daiquiris' can appropriately and literally express what I would put by "Charlie likes brunettes as people and likes daiquiris as drinks". By contrast Lyons (1977: 408) tends to think of what can be meant by the utterance as expressible by "Charlie likes$_{[people]}$ brunettes and likes$_{[drinks]}$ daiquiris", the non-parallel character of the interpretation of the utterance being determined by the different interpretations of 'like'. I think that this is a misleading way to put the interpretation.

Lyons (1977: 408) himself observes that if one thinks of the interpretation his way, the third sentence, 'Charlie likes brunettes more than daiquiris', gets interpreted as "Charlie likes$_{[people]}$ brunettes more than he likes$_{[drinks]}$ daiquiris". This interpretation seems odd because it would seem to require the same scale for people and drinks in order to make sense of 'more than', yet people-liking and drink-liking suggest different scales (Lyons 1977: 408). On my view the interpretation is better expressed by "Charlie likes brunettes as people more than he likes daiquiris as drinks", where 'more than' is not a comparison of people with drinks but a comparison of the ratio of a segment of the people-scale to its whole length with a ratio of a segment of the drink-scale to its whole length, i.e. a comparison of relative positions of brunettes and daiquiris on their respective scales or in their respective orderings. This seems perfectly coherent and not odd at all.

Lyons's (1977: 408) linguistic intuitions guide him rightly, nevertheless, for he eventually decides, for reasons not clear to me from his text, that sentence (*c*) is really not worse than sentences (*a*) and (*b*). At least, where the first two are felicitous so is the third. And he agrees that the first two can be felicitous utterances. The theoretical issue is whether the crossed interpretation is a possible literal meaning of sentence (*a*). Both Lyons (1977: 408) and I believe that it is. Lyons then goes on to assert that this suggests sense-generality, not polysemy, in 'like'. He adopts criterion (B).

Having done so, and concluded that 'like' is not polysemous, Lyons (1977: 408) retreats from his conclusion when he remarks that:

> the whole procedure is . . . of doubtful validity. Once we get to the point of convincing ourselves that, with a little imagination, we can interpret utterances like [*Charlie likes brunettes and daiquiris, Charlie likes brunettes and Jay daiquiris, Charlie likes brunettes more than daiquiris*], it is easy to start doing the same with utterances like *Mary and Ruth were*

both crying: one was weeping profusely and the other was screaming blue murder.

I do not wish to join Lyons in this retreat. What Lyons thinks is easy to start doing strikes me as not easy in the least. I cannot do it at all. I do not believe that any speaker of English can. Lyons's doubt seems to me just linguistically unavailable to speakers of English, so that Lyons's caution on this point is misplaced. Reduced co-ordinate expressions provide us with reliable criteria for distinguishing ambiguity from sense-generality. The adequacy of (A) and (B) is surely not to be doubted on the grounds that Lyons suggests.

In sum, in this section I have rejected Quine's extensionalism. It does not provide a theoretically adequate version of ambiguity. Quine's extensionalist notion of referential generality is a trivial concept. I have then expounded Lyons's reduced co-ordinate expression criteria for distinguishing ambiguity from sense-generality and rejected Lyons's scepticism about his criteria (A) for ambiguity and (B) for sense-generality.

2. Ambiguity Tests

It is sometimes forgotten that in 1967, in his Beckman Lectures "Language and Mind" at the University of California, Berkeley, Noam Chomsky pointed out that a necessary condition of deletion was sameness of sense of the deleted element with a formally identical element in the sentence. (This was probably forgotten because Chomsky immediately launches a discussion in terms of transformations requiring reference to the history of a derivation, the Chomskyan Gambit of turning a semantic observation into a syntactic problem. Too much Proof Theory is bad for one's philosophical health.) Chomsky (1972: 33) wrote:

8 I don't like John's cooking any more than Bill's cooking.
9 I don't like John's cooking any more than Bill's.
Sentence *9* is ambiguous. It can mean either that I don't like the fact that John cooks any more than I like the fact that Bill cooks, or that I don't like the quality of John's cooking any more than I like the quality of Bill's cooking.* However, it cannot mean that I don't like the quality of John's cooking any more than I like the fact that Bill cooks, or conversely, with "fact" and "quality" interchanged. That is, in the underlying structure

[associated with] (*8*) we must understand the ambiguous phrases "John's cooking" and "Bill's cooking" in the same way if we are able to delete "cooking".

* There may also be other interpretations, based on other ambiguities in the structure "John's cooking"—specifically the cannibalistic interpretation and the interpretation of "cooking" as "that which is cooked".

It is characteristic of an ambiguity of this type that *9* is two-ways rather than *per impossibile* four-ways ambiguous. Crossed interpretations are not possible as literal meanings of the sentence. The reduced form allows only parallel interpretations as possible meanings.

In an influential note on ambiguity, George Lakoff (1970: 357–8) made similar observations for ambiguous sentences:

(1′) *Selma likes visiting relatives and so does Sam.*
(2′) *Harry was disturbed by the shooting of the hunters and so was Al.*
(3′) *The chickens are ready to eat and so are the children.*

These sentences are two-ways ambiguous, not four-ways ambiguous. For example, (1′) can mean either

Semla likes going to visit relatives and Sam also likes going to visit relatives.

or

Selma likes relatives who are visiting and Sam also likes relatives who are visiting.

However (1′) cannot mean

Selma likes going to visit relatives and Sam likes relatives who are visiting.

nor

Selma likes relatives who are visiting and Sam likes going to visit relatives.

Such cases are discussed at length in Lakoff (1966), where it is claimed that identity of underlying and not superficial structure is required for the operation of the rule of VP-deletion. Similarly, (2′) and (3′) are two-ways, not four-ways, ambiguous. The same meanings are required on the left-hand and right-hand sides of (1′)–(3′).

Lakoff then applies the conjunction-reduction test for ambiguity to three controversial cases: (i) 'John hit the wall', (ii) 'John knocked the child over', (iii) 'John cut his arm with a knife', where what is in question is whether these sentences are ambiguous between "purposive" and "accidental" readings of the verb phrases. The non-existence of crossed interpretations as possible literal meanings of reduced conjunctions convinces Lakoff that (i)–(iii) are ambiguous sentences.[6] Thus Lakoff adopts criterion (A) of Chapter 2, Section 1:

(A) *The impossibility of a crossed, literal paraphrase for a conjunction-reduced sentence S entails the ambiguity of S.* (The distinct, parallel paraphrases express distinct senses.)

Following Chomsky (1957: 35–6) and Grinder and Postal (1971: 269), Zwicky and Sadock (1975) illustrate the way conjunction reduction requires identity of sense (not identity of reference); they also adopt criterion (A). Zwicky and Sadock (1975: 18) write:

> If:
> (59) *Morton tossed down his lunch.*
> were unspecified (rather than ambiguous) as to whether Morton bolted his lunch or threw it to the ground, then the parallel example:
> (60) *Oliver tossed down his lunch.*
> would also be unspecified, and the reduced sentence:
> (61) *Morton and Oliver tossed down their lunches.*
> would have four understandings, not two, because the identity condition on conjunction reduction cannot require identity of elements that are not part of syntactic structure. But (61) lacks the crossed understandings (except as a joke), and we conclude that (59) is ambiguous. To support our intuition that (61) lacks the crossed understandings, we can add contextual information so as to favour different understandings in the two predicates:
> (62) *?Morton, as always a greedy eater, and Oliver, who continued to refuse food on principle, tossed down their lunches.*

I wish to make two remarks about this passage. Zwicky and Sadock's (1975: 3 n. 9) term 'understanding' is intended by them to

cover both those elements of 'meaning' (in a broad sense) that get coded in semantic representations and those that do not. Each understanding

[6] I disagree with George Lakoff on some of these examples, e.g. 'cut'. A crossed, literal interpretation is available for 'John cut the chicken and his thumb', which suggests that 'cut' is not polysemous between "purposive" and "accidental" senses. 'John hit Brian and the highway divider' does not so obviously subsume a crossed interpretation under a literal meaning. I am willing to be pushed to the view that 'hit' is polysemous. But the differences among 'John cut Brian's flowers', 'John cut Brian's hair', 'John cut Brian's cake', and 'John cut Brian's arm' do not prove to me that 'cut' is polysemous. (For a similar view of 'cut' see Weydt (1973); for contrary views see J. F. Ross (1981).) 'Enraged, Brian knocked over his wife and also an innocent bystander' seems to me more like 'cut' than like 'hit'. See n. 17. (I am indebted to discussion with Rogers Albritton and with Rick Tibbetts.)

corresponds to a class of contexts in which a linguistic expression is APPROPRIATE [*my emphasis*]—though, of course, a class of contexts might correspond to several understandings, as in examples like *Someone is renting the house* (courtesy of J. L. Morgan).

It is noteworthy that Zwicky and Sadock associate an understanding of sentence with a class of contexts in which its utterance is appropriate, as contrasted, for example, with contexts in which its utterance is true. I shall have occasion to return to Zwicky and Sadock's comments on appropriateness and truth later in this Section and in Chapter 3, Section 1. A second matter is Zwicky and Sadock's (1975: 7 n. 17) use of the question mark before sentences. They comment that "the question mark . . . indicates an anomaly other than ungrammaticality, in particular internal contradiction or inappropriateness". The anomaly in (62) is evidence that (62) lacks the crossed understandings.

In another example Zwicky and Sadock (1975: 22–3) appeal to criterion (B) to demonstrate non-ambiguity:

(B) *The possibility of a crossed, literal paraphrase for a conjunction-reduced sentence S entails the non-ambiguity of S.* (The distinct, parallel paraphrases do not express distinct senses.)

Zwicky and Sadock (1975: 22–3) write:

. . . We can now return to examples
(6) *Melvin became as tall as any of his cousins.*
(7) *Melvin became taller than the average Ohioan.*
(8) *Melvin became the tallest linguist in America.*
and show that they exhibit no ambiguity with respect to whether Melvin or his circumstances change. The reduced sentences:
(87) *Melvin became as tall as any of his cousins, and then the same thing happened to Martin.*
(88) *Melvin became taller than the average Ohioan, and then the same thing happened to Mervyn.*
(89) *Melvin became the tallest linguist in America, and the next year the same thing happened to Merton.*
all permit the crossed understandings.

In another application of (B) as well as (A), Zwicky and Sadock (1975: 23) write:

David Dowty, Larry Martin, and Carlota Smith have suggested to us that identity tests indicate that (90) is unspecified and (91) ambiguous
(90) *John and Martha left.*

(91) *John and Martha are married.*
[between sentential and phrasal conjunction; the phrasal interpretations are 'John left (together) with Martha' and 'John is married to Martha'.]
According to them, (94) allows the crossed understandings and (95) does not.

(94) *John and Martha left, and so did Dick and Pat.*
(95) *John and Martha are married, and so are Dick and Pat.*

In the examples described in this section and the last I have shown how Lyons (1977: 408) and Zwicky and Sadock (1975: 18, 22–3) employ criterion (A) for ambiguity and criterion (B) for sense-generality.

Zwicky and Sadock (1975: 23) also introduce a qualification on identity tests that it is important to discuss. They write of the example just discussed:

> But it seems to us that the crossed understandings are available for (95) as well, since anyone who is married is married to someone.

I do not myself see how an analytic truth like 'Anyone who is married is married to someone' is evidence for the possibility of a crossed interpretation being literally and non-anomalously meant by sentence (95). Consider, for example, *?John and Martha are married, but they like each other more than they like their spouses, and so are Dick and Pat, who are as romantic about each other now as they were when they exchanged their vows*. By Zwicky and Sadock's own showing this sentence, which I find anomalous, is evidence against the availability of a crossed interpretation of (95). So I find myself in disagreement with their judgement about (95). They go on to say:

> Similarly, it is not surprising that (94) can have crossed understandings, since anyone who has left together with someone has left, and the fact that someone has left does not exclude the possibility that he left together with someone.

Again, why should trivial remarks about entailment and logical consistency be grounds for a claim about the possibility of (94)'s having a crossed understanding? (An understanding, on Zwicky and Sadock's (1975: 3 n. 9) own view, is a meaning (in a broad sense) of a sentence.) In the case of (94) we can expand it to *John and Martha left, but a half-hour apart, and so did Dick and Pat, who never go anywhere separately*. This sentence is not anomalous, which supports the sense-generality of (94). So, what bearing do the logical relations of entailment and consistency have on the permissible meanings of (94)?

To understand Zwicky and Sadock's view of the matter I must introduce their notion of *privative opposition*. They (1975: 6) note that 'parent' and 'mother' have understandings that are privative opposites with respect to gender. The gender feature is a binary semantic feature (Zwicky and Sadock 1975: 6 n. 14). In particular, 'mother' and 'parent' are privative opposites with respect to gender if 'mother' can be represented as being identical to 'parent' except that 'mother' includes some specification for gender, e.g. [– MALE], that is lacking in 'parent'.[7] Zwicky and Sadock (1975: 6–7) go on to remark that "one privative opposite (the more specific understanding) implies the other (the more general understanding)". Zwicky and Sadock assume that the concept of privative opposition applies to the understanding of sentences as well as to the understanding of lexemes. Furthermore, they believe that the understanding "John and Martha left (together)" is the privative opposite of "John left and Martha left".

In my view the extension of the concept of privative opposition to sentences requires some defence, which Zwicky and Sadock do not provide. The extension can be made (I extend the notion of sense-generality to sentences from its paradigm applications to General terms), but it should not be made so unselfconsciously. It is also clear that Zwicky and Sadock assume that the entailment of 'John left and Martha left' by 'John and Martha left (together)', and the failure of the converse entailment, justifies the description of the second sentence as the privative opposite of the first. That is a gratuitous assumption. Privatively opposite lexemes will show such entailment and failure of entailment relations, but it surely does not follow that any expressions satisfying such relations are privative opposites. Otherwise 'The moon is made of green cheese' is the privative opposite of '2 + 2 = 4', and 'Charlie is handsome' is the privative opposite of 'Charlie is handsome or Charlie is witty'. Those consequences strike me as so peculiar that it is completely unclear what theoretical work Zwicky and

[7] Incidentally, this is not how Lyons (1977: 279) defines the notion. He writes, "A privative opposition is a contrastive relation between two lexemes, one of which denotes some positive property and the other of which denotes the absence of that property: e.g., 'animate': 'inanimate'."

Sadock's concept of *privative opposition* is now supposed to be doing. The only close parallel to the lexical paradigm of privatives would be the entailments $A\&B \Vdash A$ and $A\&B \Vdash B$.

Zwicky and Sadock (1975: 23) go on to say of (94) and (95) that:

> We are dealing here with privative oppositions, so that no matter what the linguistic state of affairs, by applying identity tests we will always conclude that we are dealing with a lack of specification; the existence of the more general understanding guarantees that we will get all possible understandings.

The word 'get' in their last clause strikes me as ambiguous (general?). Do they mean that the sentence (95) can mean, i.e. be literally paraphrased by, parallel and crossed interpretations? Or, do they mean that (95), in its parallel, "more general" interpretation, is logically consistent with two crossed propositions? Their ARGUMENT suggests the latter; their INTENTIONS suggest the former. In fact, I believe that they have conflated the two. They believe that the latter guarantees the former, but it does not.

The logical consistency of $A'\&B$ with $A\&B$, where $A \Vdash A'$, hardly implies that $A\&B$ is an interpretation of $A'\&B$, i.e. that $A'\&B$ can MEAN $A\&B$. The reasons Zwicky and Sadock give for being unsurprised by the existence of crossed understandings for (94) are very dubious.

As for (95), their argument has the form: since $A\&B$ entails $A'\&B$ and $A\&B'$, where $A \Vdash A'$ and $B \Vdash B'$, $A\&B$ can MEAN $A'\&B$ (i.e. $A\&B$ has $A'\&B$ as a possible, literal interpretation), and $A\&B$ can MEAN $A\&B'$. This view is at best the old mistake of conflating verifications with meanings (see Chapter 2, Sections 1 and 3).

Zwicky and Sadock (1975: 3) have connected their notion of understanding (meaning in the broad sense) with a notion of appropriate (non-anomalous) utterance in a context. But they say of *privative oppositions* to which conjunction reduction applies that:

> No appeal to contexts will help, because every context in which the crossed understandings . . . are *appropriate* is also a context in which the [more general] parallel understanding is *appropriate*. Therefore, we cannot test the possible understandings . . . by supplying a context that forces one of the crossed understandings, as we did in the discussion of (61)–(62) . . .

(61) *Morton and Oliver tossed down their lunches.*
(62) *?Morton, as always a greedy eater, and Oliver, who continued to refuse food on principle, tossed down their lunches*;

if we *eliminate* the [more general] parallel understanding[∅], we also *eliminate* the crossed understandings.

This passage becomes true only if one substitutes 'true' for 'appropriate' and 'falsify' for 'eliminate'. Appropriateness is a matter of meaning, not a matter of truth-conditions or logical equivalences. It is perfectly clear that a crossed understanding $A\&B'$ entails $A'\&B'$, where $A \Vdash A'$ and $B \Vdash B'$, but NOT every context in which $A\&B'$ is appropriate will be a context in which the logically weaker proposition $A'\&B'$ is appropriate, if by 'appropriate' one means 'felicitous' or 'relevant'. If Zwicky and Sadock by 'appropriate' merely mean 'grammatical' or 'semantically consistent', the relativity to context becomes vacuous, and their claim becomes true but empty.

Zwicky and Sadock conclude that an identity test is not decisive for determining the ambiguity of privative opposites. I believe that their arguments fail to establish their conclusion. I also believe that their notion of a SENTENTIAL privative opposition of sentences A and B is nothing more than (i) A entails B (i.e. $A \Vdash B$), and (ii) B does not entail A (i.e. $B \nVdash A$). This is a matter of truth in a model, as matters in formal semantics are, but not essentially a matter of linguistic meaning. The terminology 'privative opposition', borrowed from lexical semantics, camouflages Zwicky and Sadock's conflation of truth-conditions with sentential meaning.[8]

One textual indication of their conflation of sense and reference is found in the following discussion (Zwicky and Sadock 1975: 6):

Thanks to the fact that English distinguishes a set of lexical items that are masculine (*man*, *boy*, *king*, etc.) from corresponding items that are feminine (*woman*, *girl*, *queen*, etc.) and the fact that English pronominal reference systematically distinguishes between masculine and feminine, the differentia masculine/feminine plays a part in the semantic system of English.

So far so good, except that the description of English pronouns in terms of their reference rather than their sense suggests a confusion of a semantic property of expressions, whether "male"

[8] See Atlas (1977*b*: 328–30), Roberts (1984, 1987), Sadock and Zwicky (1987), and Chapter 3, Section 1.

or "female" is part of the sense, which we could describe as the (semantic) gender of a lexical item, with a biological property of the denotata of such expressions, the sex of an individual. Zwicky and Sadock continue:

> But from this we can conclude nothing about the status of lexical items like *person*, *actor*, *chairman*, *secretary*, *dog*, or *goose*, all of which can be understood as either masculine or feminine.

An individual denoted by 'person' might be expected to be, or have been, either male or female (except for instances of hermaphroditism). But these expressions are not "understood" as either masculine or feminine: it is not the case that particular utterances are "understood" as masculine, while other utterances are "understood" as feminine. The utterances are not even "understood" as masculine-or-feminine. These words are neither marked as masculine nor marked as feminine. They are also not marked as masculine-or-feminine. They are semantically unmarked; the sex of the referent is not lexicalized in the sense of the expression. They are understood as neither masculine nor feminine, i.e. '*x* is a person' does not entail '*x* is male', and '*x* is a person' does not entail '*x* is female'. These expressions are sense-non-specific for gender. What can be understood is that such expressions have extensions whose members are male, or female, or hermaphroditic, etc. The referent will be understood to be either male or female, etc; the sense will be understood to imply neither maleness nor femaleness of any referent.

This sense-non-specification is to be distinguished from reference-non-specification. Zwicky and Sadock (1975: 2–3) illustrate *reference-non-specification* with the example *My sister is the Ruritanian Secretary of State*, which they say is:

> unspecified (general, indefinite, unmarked, indeterminate, vague, neutral) with respect to whether my sister is older or younger than I am, whether she acceded to her post recently or some time ago, whether the post is hers by birth or by merit, whether it has an indefinite tenure or will cease at some specific future time, whether she is right-handed or left-handed, and so on.

These respects in which the sentence is referentially unspecified are just the unboundedly diverse aspects of those possible states of affairs that would verify the sentence. It should be obvious that these respects are not at all the respect in which, e.g., the words 'mother', 'father', and 'person' are either specified or unspecified for gender, viz. having or lacking a semantic gender feature, as Zwicky and Sadock (1975: 6) themselves note. Yet they fail to attend to the distinction between sense-non-specification and reference-non-specification.

In this section and the last I have expounded and criticized some standard and focal linguistic discussions on the subject of ambiguity. The linguists, following Chomsky's (1957: 35–6) lead, have proposed useful criteria (A) and (B) for distinguishing ambiguity from sense-generality.[9] I have focused my criticism of Lyons (1977) and of Zwicky and Sadock (1975) on their discussions of a conjunction-reduction identity test because the test has antecedents in the work of Gilbert Ryle (1949) and of Aristotle, and because it raises fundamental issues of sense and reference, truth and meaning. I have criticized Lyons for being misled by his erroneous interpretations of sentences into making pseudo-objections against his own criterion (B). I have criticized Zwicky and Sadock for their mistaken analyses of sentences (94) and (95), their conflating the question of literal paraphrase with the question of logical consistency, their ignoring the distinction between reference-non-specification and sense-non-specification, and in general their not keeping straight whether they are talking about words or about objects.

3. The Erroneous Disjunction View of Generality of Sense[10]

In a recent criticism of Israel Scheffler's (1979) discussion of ambiguity, Avishai Margalit (1983) distinguishes ambiguity from sense-generality by appealing to linguistic tests. In this section I shall discuss one of these tests, the "known disjunction test", as Margalit puts it, which Margalit has adopted in criticism of Scheffler. I shall argue that Margalit's "disjunction test" of sense-generality is not well-defined and fails to save the linguistic phenomena, begs the explanatory questions, and changes the linguistic subject rather than explicates the linguistic doctrine of

[9] For a characteristic use of these criteria, see E. Ejerhed (1981: 240–3) in Heny (1981).

[10] Chapter 2, Section 3 includes a revised version of Atlas (1984*b*), © 1984, by D. Reidel Publishing Co.

sense-generality. Because of the interesting philosophical use to be made of this concept, it is important to understand its character. It is important because the distinction has been used to evaluate and revise conventional analyses of negation (Atlas (1975*b*, 1977*b*, 1979), Gazdar (1979, 1980), Horn (1978*a*), Kempson (1975)), intensional contexts (Kempson 1979), adverbial modification, quantifiers (Kempson and Cormack (1981), Bach (1982), Sadock (1975)), truth-value gaps (Atlas 1975*b*, 1977*b*), and relevance logic (Lewis 1982) as well as to criticize theories of meaning modelled on Tarski's theory of truth or on possible-world semantics (Atlas (1975*a*, 1978), Partee (1979, 1981)). Whenever philosophers have concerned themselves with possible ambiguity, e.g. 'exists' in 'Prime numbers exist' and 'Aardvarks exist', 'if' in 'I would have done otherwise if I had wished' and 'The match would have lit if it had been struck', 'know' in 'I know that Rogers Albritton lives in Los Angeles' and 'I know how to ride Brian Maddox's bike', there has followed a drive toward parsimony in the positing of meanings, notably in the work of Morton G. White (1956), W. V. O Quine (1960), and G. E. M. Anscombe (1981).

Margalit observes that

(14) *Jim did not play any game, he played chess.*

is false, but

(15) *Jim did not see any pitcher, he saw a vase.*

can be true. This means, Margalit believes, that 'game' is sense-unspecified with respect to the lexical feature by virtue of which it would distinguish chess from baseball, from tennis, etc., and that 'pitcher' is ambiguous, i.e. has two meanings, one by virtue of which the expression designates a baseball player, the other a vessel for pouring out liquid. His justification for the interpretation of the linguistic data is this (Margalit 1983: 132):

> The idea is to view a general term as a disjunctive term, so that its negation, in an appropriate sentence, negates all its disjuncts as well as their entailments. With an ambiguous term, on the other hand, the negation holds for one of its meanings but not necessarily for all, so that one can add a conjunct that is analytically entailed by the term without rendering the sentence false. This test is of course based on the notion of analytic entailment ('if x is a pitcher, x is a vase') . . .

I should dispose of the trivial points first. Suppose we do treat a general term like 'game' as a disjunction, so that 'x is a game' is understood as 'x is chess or x is baseball or . . . or x is pick-up-sticks'. *There are baseball games* entails *There are baseballs*, which entails *There are balls*, but if I denied that there are baseball games I would not logically imply that there are no balls. So when Margalit writes that the negation of a general term negates the entailments of the disjuncts into which the term has been analysed, Margalit has formulated the point carelessly. In addition, Margalit should have written: 'one can add a conjunct that analytically entails the term that is negated without rendering the sentence false. This test is of course based on the notion of an analytic entailment ("if x is a vase, x is a pitcher") . . .'.

There are two major issues. The first is how Margalit interprets his linguistic observations, and the second is the theory that supports that interpretation. I wish to separate the arguments involved in Margalit's interpretation of the linguistic data from further theoretical propositions involved in the justification of that interpretation. The linguistic arguments about the data do not support the theory that supposedly explains them. Furthermore, I wish to show on independent grounds that the theory is implausible. First, I consider the linguistic arguments about the data.

Margalit's test has two parts, one for sense-generality, one for ambiguity. First, Margalit hopes to persuade us that where a sentence has the form $\sim Fa$ & Ga, and where $Ga \Vdash Fa$, if the sentence is false (in fact, analytically false), F is sense-unspecified for some lexical feature. Second, if the sentence can be true, it can be true by virtue of some disambiguation F' of F, or G' or G, one in which $Ga \Vdash F'a$ is not true, or $G'a \Vdash Fa$ is not true, and so F, or G, is ambiguous.

Let us compare with Margalit's examples (14) *Jim did not play any game, he played chess* and (15) *Jim did not see any pitcher, he saw a vase* the following sentence:

(16) Jim did not cook eggs, he cooked broccoli and eggs.

and ask Margalit's question: are statements of (16) true or false? I shall argue that, so far as intuitions about the statement's being true or false are concerned, there are intuitive grounds for taking some statements of (16) to be true and some statements of (16) to be false. It would then follow, by Margalit's reasoning, that 'eggs' is both ambiguous and general, so that on Margalit's interpretation the test is not well-defined.

If one is rather literal-minded, one would take a statement of

(16) to be false. Since '*x* cooked broccoli and eggs' entails '*x* cooked eggs', if one conjoins the negation of the latter with the former, the resulting conjunction is false, in fact necessarily false. To Margalit this would mean that 'eggs' is unspecified with respect to the lexical features by virtue of which it would distinguish broccoli and eggs from ham and eggs, etc., so that according to Margalit's reasoning we should analyse 'eggs' as, in effect, a disjunction 'eggs with broccoli or eggs with something different from broccoli'. A logician, or a philosopher committed to extensionality, might be tempted by the idea that '*x* is a game' means '*x* is chess or *x* is baseball or . . . or *x* is pick-up-sticks'. By coincidence, at some time the expressions might be extensionally equivalent, but of course it is utterly implausible that a lexicological theory would correctly characterize the meaning of 'eggs' by 'eggs with broccoli or eggs with something different from broccoli'.

For those with an ear for English, another understanding of sentence (16) is available. We can easily imagine discourse in which the statement would be understood as true, e.g. with contrastive stress, *Jim didn't cook* EGGS, *he cooked* BROCCOLI *and eggs*, which one could approximate by the cleft sentences, 'It was not eggs that Jim cooked, it was broccoli and eggs that he cooked'.[11]

For the nonce let us hold to the intuition that some statement of sentence (16) could be true. This would mean to Margalit that 'eggs' is ambiguous. It might be, for example, that 'eggs_1' means "eggs only"; 'eggs_2' means "eggs with something (else)"; what would be communicated in the statement would be the true proposition "Jim did not cook eggs (only); he cooked broccoli and eggs (with something (else))". Nothing in Margalit's treatment of the data blocks the conclusion that 'eggs' is ambiguous.

But there is a further complication. If 'eggs' were ambiguous between 'eggs_1' and 'eggs_2' its meanings would be distinguished by expressions one of which, 'eggs with something', looks itself to be "general" in the disjunctive sense that Margalit adopts, e.g. 'eggs with *u*, *v*, or *w*'. We could avoid this curious complication if the truth of a statement of (16) were explained by a different ambiguity: 'eggs' means 'eggs_1', i.e. "eggs only", and 'eggs_3', i.e. "eggs with broccoli". But in that case, if true, the truth of a

[11] For various reasons this paraphrase is only approximate; it is not synonymy; see Atlas and Levinson (1981).

statement of (16) is due to the truth of the proposition "Jim did not cook eggs (only); he cooked broccoli and eggs (with broccoli)". Unfortunately, this analysis and similar analyses of the ambiguity of 'eggs' seem just as far-fetched as the one creating a "general" sense of an ambiguous expression.

By contrast with Margalit's examples where 'game' is shown to be general by example (14) and 'pitcher' is shown to be ambiguous by example (15), sentence (16) shows 'eggs' to be both general and ambiguous. In that case the test, which is supposed to distinguish the two cases, would hardly be adequate. On Margalit's interpretation the test is not well-defined.

The 'broccoli and eggs' example shows that something more than intuitions about a statement's falsehood or truth and the success or failure of an entailment is required for coherently distinguishing sense-generality from ambiguity. The obvious way to free Margalit from his difficulties with ambiguity is to note that on the understanding of (16) in which it is true, asserting *Jim didn't cook EGGS* implicates "Jim cooked something" (Atlas and Levinson 1981: 52).[12] Asserting *Jim didn't cook EGGS* also implicates "Jim cooked eggs but not just eggs" (Grice 1978: 122–3), which is consistent with *Jim cooked BROCCOLI and eggs.*[13] Omitted from Margalit's test are criteria for distinguishing ambiguity from univocality cum generalized conversational implicata (and the contents of "indirect speech acts") in the case in which some statement of the sentence is true. Coherent application of the test requires such criteria (Sadock 1978).

Whether the understanding of (16) in which it is false is grounds for the generality of 'eggs' depends on the correctness of the interpretation of the concept of generality in the linguistic test. For example, in Margalit's interpretation of the test sense-generality is modelled in classical logic by a disjunction. Something more than intuitions about a statement's truth-value and facts about entailment

[12] Grice (1978: 123) claims, in effect, that the implicatum is "Jim cooked something (other than eggs)". He does not distinguish between the presuppositional implicatum "Jim cooked something" and the implicatum "Jim cooked something other than eggs". Nor does he notice that 'other than eggs' is ambiguous between "different from eggs" and "in addition to eggs". From the context it is clear that he intends the former.

[13] For discussion of the implicatum in *Jim didn't cook EGGS*, see Levinson (1983: 139 n. 25), Atlas (1980*a*), Horn (1978*a*: 136–7), Grice (1978: 122–3). Steven Boër (1979) raises important issues that for present purposes I ignore.

is required for distinguishing ambiguity from Margalit's disjunction conception of general univocality, because whether or not the analytic falsehood of (14) *Jim did not play any game, he played chess* warrants the univocality of 'game', it does not entail that '*x* is a game' MEANS "*x* is chess or *x* is baseball or . . . or *x* is pick-up-sticks". So I now turn to Margalit's justification of his test for generality, the disjunction view of generality.

The motive for a lexicological distinction between ambiguity and sense-generality lies in the desire to describe, e.g., the difference between pronouns that are specified for gender, like 'he' and 'she', and nouns that are not, like 'neighbour'. One would normally think that '*x* is a neighbour' means "*x* dwells nearby". The sense-non-specification of 'neighbour' with respect to gender means that the lexical entry for 'neighbour' contains no semantic marker for gender; 'neighbour' is unmarked in that respect. Gender is irrelevant to its meaning.

It is sometimes suggested that this non-specification implies that '*x* is a neighbour' means "*x* is a nearby-dweller, and *x* is male or *x* is female". This is the disjunction view that Margalit adopts (see Kempson (1975: 15–16), Hollings (1980), Kempson and Cormack (1981: 262–3)). I wish to show the difficulties this conception leads to. It is a mistaken view of sense-generality. I shall discuss the 'neighbour' example, since it is agreed to be a typical example of generality.

Those who accept the disjunction picture of generality are committed to, and in fact accept, the view that their analysis '*x* is a nearby dweller, and *x* is male or *x* is female' is equivalent to '*x* is a nearby dwelling male or *x* is a nearby dwelling female'. Then, '*x* is not a neighbour' means "*x* is neither a nearby dwelling male nor a nearby dwelling female" (Kempson (1975: 11–13), Kempson (1979: 289)).

I wish to upset the pleasant tidiness of this picture by considering several cases. Suppose the Archangel Gabriel lives in an apartment above mine. On the disjunction view my statement *Gabriel is not my neighbour* would be true, because of the sexual condition of angels, rather than false, because of the housing conditions of this angel. Or, suppose Brunhild, a person with XXY sex-chromosomes, like some East German "women" athletes, lives in an apartment adjacent to mine. On the disjunction view my statement *Brunhild is not my neighbour* would be true, because of what Brunhild is like physically, rather than false, because of where Brunhild lives.

There are more, and more controversial, examples. To these cases of individuals that are neither male nor female George Lakoff would add transsexuals and hermaphrodites (Lakoff 1982: 26). Even in our ordinary conceptual scheme hermaphrodites are both male and female rather than neither male nor female, so I reject that example. For the sake of argument, suppose Lakoff is correct that transsexuals are neither male nor female, in the ordinary employment of 'male' and 'female' in English. Now, Renée Richards and Jan Morris are not my neighbours, but they could have been. No empirically adequate semantic theory of 'neighbour', 'male', and 'female' can rule out the possibility of so describing them. On the disjunction picture, however, even were Richards and Morris living in apartments on each side of mine neither could be truly said to be my neighbour. In those circumstances the disjunction picture would make *Jan Morris is not my neighbour* true. I should be forced to invent a word 'schneighbour' such that *Jan Morris is my schneighbour* is true; 'x is a schneighbour' means "x dwells nearby".

What distinguishes 'schneighbour' from 'neighbour', as the latter is analysed in the disjunction picture, is precisely that 'schneighbour' is sense-unspecified for gender, so that I have a word to describe my angelic, XXY, or transsexual neighbours, should I have any.[14] Of course, this is just what my ordinary English word 'neighbour' actually does. The disjunction characterization of the non-specification of 'neighbour' with respect to gender fails to account for a straightforward employment of 'neighbour' in English. But the disjunction theorist has two replies, my disagreement with which brings out further difficulties with the disjunction view of generality.

First reply: "Your examples merely show that 'x is a neighbour' should be analysed as 'x dwells nearby, and x is male or x is female or x is an XXY or x is a transsexual or x is an angel'." Response:

[14] Of course I mean 'general' in the sense explained five paragraphs above, not in the mistaken explication by disjunction (D), according to which 'x is a schneighbour$_D$' would mean "x dwells nearby, and x is male or x is female". For, then, 'x is not my schneighbour$_D$' would be open to the very objections just posed to 'x is not my neighbour$_D$'. One would then have to introduce another predicate 'schneighbourle' that was semantically unspecified with respect to gender, in the sense I intend, in order to express what the English word 'neighbour' means.

This suggestion not only shows a monomaniacal commitment to a picture but also an inability to recognize *ad hoc* hypotheses, and shows a naïve understanding of the point of a semantic theory. I am not interested in historical accidents of coextensiveness between predicates in our language at some time. I am not even interested in the encyclopaedic view of God's Anthropologist, who knows how to fill out the truth-conditions for all times, places, and values of the individual variable in '*x* is a neighbour' or '*x* is a game'. I am interested in characterizing those properties of the meaning of an expression that will explain how it manages to perform the designating function that it actually does perform in our use of the language. The reply amounts to saying that 'neighbour' severally designates males, females, XXYs, transsexuals, angels, *et al.* because it severally designates males, females, XXYs, transsexuals, angels, *et al.* It is an explanatorily vacuous proposal.

Second reply: "Your examples are compelling, but only because your original target was a straw man. This disjunction theorist should have proposed ''*x* is a neighbour' means "*x* dwells nearby, and *x* is male or *x* is not male"' rather than ''*x* is a neighbour' means "*x* dwells nearby, and *x* is male or *x* is female"'. If one were to accept this revised analysis, the difficulties you pose would not arise." Response: Would the disjunction theorist propose ''*x* is a game' means "*x* is chess or *x* is not chess"'? No, that would be silly, but no sillier than the revised disjunction analysis. One might just as well have proposed ''*x* is a neighbour' means "*x* dwells nearby, and *x* is female or *x* is not female"', or ''*x* is a neighbour' means "*x* dwells nearby, and *x* has Down's Syndrome or *x* does not have Down's Syndrome"', or ''*x* is a neighbour' means "*x* dwells nearby, and *x* was born 19 May 1958 or *x* was not born 19 May 1958"'. From an extensional point of view, the difference between these proposals is nil. From an intensional point of view, it is important to recall the original goal: distinguishing the non-specification of 'neighbour' from the specification of, e.g., 'he' and 'she' with respect to gender. In grammatical theory gender is an "entrenched" predicate (Goodman 1983); non-maleness, the complement of maleness, is not a gender.[15] To say that 'neighbour' is unspecified for gender is NOT to say that 'neighbour' is, so to

[15] Nor are predicates complementary to projectible predicates typically projectible. See Foster (1971) and Quine (1969).

speak, tautologously specified for gender, e.g. '*x* dwells nearby, and *x* is male or *x* is not male'.

The error in the disjunction picture of generality arises from the following fallacious argument: '*x* is a neighbour' means (denotes what is denoted by) '*x* is a nearby dwelling male' or '*x* is a nearby dwelling female'. So, '*x* is a neighbour' means '*x* is a nearby dwelling male or *x* is a nearby dwelling female'. In other words, '*x* is a neighbour' means '*x* is a nearby-dweller, and *x* is male or *x* is female'. Therefore '*x* is a neighbour' means (is synonymous with) '*x* is a nearby-dweller, and *x* is male or *x* is female' (Kempson (1975: 16), Kempson and Cormack (1981: 262–3)).

In the case of an ambiguous expression, e.g. 'bachelor', the confusion in the argument is more strikingly exhibited: '*x* is a bachelor' denotes what is denoted by '*x* is a young knight' or by '*x* is an adult, unmarried male'. So, '*x* is a bachelor' denotes what is denoted by '*x* is a young knight or *x* is an adult, unmarried male'. Therefore '*x* is a bachelor' is synonymous with '*x* is a young knight or *x* is an adult, unmarried male'.

These arguments commit the fallacy of equivocation, on 'means', and confuse the use and mention of expressions: 'means "*A*" or "*B*"' is not the same as 'means "*A* or *B*"'. Because of the character of sense-general terms, the confusion is merely less obvious in the first argument than in the argument with an ambiguous term.[16]

Margalit's deployment of linguistic tests distinguishing ambiguity from generality is a significant advance beyond the naked intuitions about difference in denotation that Scheffler (1979: 13–15) perforce must appeal to in drawing the distinction between an ambiguous and a univocal, General, as contrasted with Singular, term. Explanatory criteria for the distinction have been ignored by contemporary philosophers for far too long.[17] In remedying this

[16] For an independent statement of this point for ambiguous terms, see J. F. Ross (1981: 198–201).

[17] Serious contemporary linguistic discussion seems to have begun with work by Dwight Bolinger and Uriel Weinrich in the early 1960s; see Zwicky and Sadock (1975). Serious philosophical discussion began with Aristotle's *Topica*, bk. I. 15 (106–7). J. F. Ross (1981) takes seriously the various sufficient conditions for ambiguity that Aristotle suggests, and develops a full-blown defence of wide-scale polysemy in English on its basis. Of the fifteen Aristotelian conditions Ross discusses, and recommends in various forms, eleven are patently incorrect, two are circular, and four presuppose a dubious Aristotelian doctrine of categories. Not one of them survives theoretical scrutiny. Ross's book is a work whose

intellectual defect, however, Margalit unwittingly perpetuates an account of generality that distorts the concept.[18]

Margalit's interpretation of the test distinguishing ambiguity from sense-generality yields an ill-defined test. The disjunction picture fails to save the linguistic phenomena, begs the explanatory questions, and changes the linguistic subject rather than explicates the linguistic doctrine. It is important not to misunderstand the character of the concept or to underestimate the radical difference between it and the notions familiar to philosophical logicians and formal semanticists.[19] It is a Wittgensteinian, not a Carnapian, concept—but that may be merely a sociological description of the intellectual mistake that I have corrected here.[20]

It is not as if philosophers have not been warned of their misapplication of disjunction or their confusion of verifications with senses. An instructive warning was given long ago by G. E. M. Anscombe (1981: 180–95). To end this section I shall review Anscombe's analysis, hoping (no doubt vainly) finally to bury this error and not merely to dispraise it.

How would one interpret the statement *Lauren was ill after Jane was*? Anscombe suggests that it might be thought to mean:

(17) *Lauren began being ill after Jane began to be ill.*
(18) *Lauren began being ill after Jane stopped being ill.*
(19) *Lauren was ill after Jane began to be ill.*
(20) *Lauren was ill after Jane stopped being ill.*

Assuming that Jane began to be ill, (18) entails (17), and (20) entails (19). But when (19) is true, it is true whether or not Lauren began after Jane and whether or not she was ill after Jane stopped. Anyone asserting our original statement may have any of (17)–

fundamental, Aristotelian assumptions are, in my view, completely mistaken, but whose criticisms of philosophical views, and incidental linguistic observations, are correct and apt. See n. 6.

[18] Dana Scott's lessons have yet to be learned; see Scott (1973).

[19] Commentators on Wittgenstein have confused family resemblance terms, e.g. 'game', with vague terms, e.g. 'heap'. Wittgenstein was careful not to make that mistake. Family resemblance terms are "unbegrenzt" (not demarcated, *Philosophical Investigations* ¶70); vague terms are "unscharflich begrenzt" (not sharply demarcated, *Philosophical Investigations* ¶99, cf. ¶76; *Zettel* ¶392). Family resemblance terms are not vague, and theories of vague predicates do not suffice as theories of these unspecified predicates. For a typical view see Lycan (1984: 66–7), and the comments in Atlas (1988*a*).

[20] Otherwise intelligent criticism has been made irrelevant by this mistake, e.g. Hollings (1980) and Blackburn (1983).

(20) in mind, but his statement is no more ambiguous than (17) is ambiguous because someone may assert it when having in mind that Lauren began after Jane began and before she stopped, while the addressee understands it literally and so less specifically. As Anscombe (1981: 184) remarks:

> We must distinguish between a statement's covering a variety of cases and its being ambiguous. The reason we are tempted not to make this distinction here is that we are likely to be interested in which possibility is the actual one, and are quite likely to make the statement having just one of these possibilities in mind. If someone else makes the statement, and we ask which of the possibilities is the actual one, we may frame our question thus: 'What do you mean? Do you mean . . . or . . .?'

What the speaker means and what the words mean are conceptually different. And there is a logical point to this distinction. As Anscombe remarks, "That '*p* after *q*' is not ambiguous, but has alternative verifications—different *causae veritatis*—would have no importance, if we could not use '*p* after *q*' in inference without having to reason separately on each of the possible verifications." But we do reason from the literal meaning of a sentence. Anscombe continues:

> For example, we may say 'James was at large after Smith was ill; therefore he could have seen him and observed such and such effects of his illness'; [and] 'James was in touch with John after John was working on his novel, so the possibility of plagiarism cannot be ruled out'. We can draw the conclusions without knowing whether James was also at large before Smith was ill, or also in touch with John before John was working on the novel; and without knowing whether James was at large only after Smith stopped being ill, or in touch with John only after John stopped working on the novel.

Anscombe adds a further example that quite clinches the point:

> '*A* is *B*'s brother' means '*A* is male and *A* and *B* have a common parent'. It might be taken for granted that *A* was supposed to have both parents in common with *B*; or that we might be interested to inquire whether he had one or two parents in common with *B* and, if only one, whether it were father or mother; and that we might express our inquiry in the form 'what do you mean, his brother?' But this fact does not tend to show that '*A* is *B*'s brother' is ambiguous. Ambiguity is quite different: a statement is ambiguous, not when it has alternative verifications, but when it has alternative interpretations, like 'the fall of the barometer alarmed him'.

This might mean that the crash of the instrument to the ground alarmed him, or that the sinking of the mercury level in the barometer alarmed him, and no one would suggest that it is a statement explicable as a disjunction because it is verified by more than one situation.

This deserves some emphasis. No one should think that when a sentence has two or more meanings this fact should be stated: the sentence *S* means "*p* or *q*". One might say that utterances of *S* can mean that *p*, and that utterances of *S* can mean that *q*. This point about *the unacceptability of explicating an ambiguous sentence by a disjunction* of its meanings is independent of the previous point that *the alternatives in question are alternative senses, not alternative verifications*. The way Anscombe put her objections tends to conflate these two useful but separate points.

Quine does not take Anscombe's latter point because he holds, in effect, that there are nothing but verifications, and so, allegedly, no confusion between meanings and verifications can arise. Quine's abandonment of the theory of meaning for the theory of reference has had deleterious consequences for the scientific study of language. In Quine's view my taking seriously the meaning/reference distinction is akin to taking seriously the witch/human distinction. My defence, like Jerrold Katz's, is that the positing of a theoretical distinction is to be justified by the empirical fruits of the theory for our understanding of language. However interesting as philosophy, Quine's views are just too restrictive to be good science. As for Anscombe's former point, the tendency to explicate alternative senses or alternative states of affairs as a disjunction is perennial in philosophy. The naturalness of the tendency, and its sanctification by established practice, do not make it any less mistaken. If the tendency to explicate ambiguity by disjunction is both powerful and erroneous, the tendency to explicate sense-generality by disjunction has, for similar reasons, been equally attractive but, none the less, equally mistaken.

In Chapter Two I have discussed criteria for the ambiguity of expressions offered by Gilbert Ryle (1949) and W. V. O. Quine (1960). I have argued that Quine's criteria are merely an effort to change the subject of meaning to the subject of reference. Parallel to Quine's distinction between referential generality and referential "ambiguity" lies an important distinction between sense-generality and ambiguity. It is this distinction that John Lyons (1977), Chomsky (1972), G. Lakoff (1970), Weydt (1972), and Zwicky and Sadock (1975) have investigated. Philosophers of language should not ignore these investigations.[21] I have defended the co-ordinate-conjunction-reduction criteria of Lyons against his self-doubt in his own criteria, expounded, clarified, and corrected parts of the now classic discussion of ambiguity tests by Zwicky and Sadock, and refuted the analysis of sense-generality as a logical disjunction of senses or of truth-conditions. In particular:

(i) The impossibility of a crossed, literal paraphrase for a conjunction-reduced sentence *S* entails the ambiguity of *S*. (the distinct, parallel paraphrases express distinct senses, e.g. ?*John and Brian were both crying: one was weeping profusely and the other was screaming blue murder*.)

(ii) The possibility of a crossed, literal paraphrase for a conjunction-reduced sentence *S* entails the generality of *S*. (The distinct, parallel paraphrases do not express distinct senses, e.g. *Charlie likes brunettes and duiquiris, the former as people, the latter as drinks*.)

(iii) 'x is a person' does not entail 'x is male'; 'x is a person' does not entail 'x is female'. 'Person' is sense-unspecified for gender.

(iv) 'x is a person' is not synonymous with 'x is a male human $\vee$ x is a female human'. A general sense is not a disjunctive sense.

I shall now apply these ideas to a few of the most famous problems in twentieth-century analytic philosophy of language.

[21] A few philosophers have recognized the importance of the problem of ambiguity, but in the absence of a substantial, theoretical discussion, their discussions have remained preliminary: e.g. Richman (1959), Wertheimer (1972), Parsons (1973), Reeves (1975), Kripke (1977), Donnellan (1978), Cohen (1985, 1986), Roberts (1984, 1987), and Grandy (1988). This book tries to take substantial steps toward a theory. (Rogers Albritton brought Wertheimer's book to my attention after my work on my book was completed. It is a pity that Wertheimer's remarkable book has never been given the attention by philosophers that it deserves.)

3

Making New Sense of Negation, Presupposition, and Non-Existence: A Case-Study in Philosophical Linguistics

1. Presupposition and the Generality of Sense of Negative Sentences: The Atlas–Kempson Thesis[1]

There are many moth-eaten orthodoxies in the philosophy of language. In this section I intend to question one of them and to suggest an alternative view. But first, let us review the orthodoxies:

(I) English sentences of the forms 'The *F* is *G*' and 'The *F* is not *G*' have underlying logical forms. These logical forms are familiar and represented in well-known quantificational languages.

(II) 'The *F* is not *G*' is ambiguous. The surface form has two readings corresponding to two underlying logical forms.

(III) This ambiguity is attributable to the ambiguous role of the English word 'not'. Depending on the view, the ambiguity is either a structural ambiguity of scope or a lexical ambiguity disambiguated in logical form by distinct negation operators or connectives.

(IV) Since the reading of 'The *F* is *G*' and the narrow-scope or predicate-negation reading of 'The *F* is not *G*' both entail that there is a unique *F*, 'The *F* is *G*' semantically presupposes that there is a unique *F*. If the presupposition is false, the reading of 'The *F* is *G*' is neither true nor false; likewise, the predicate negation reading of 'The *F* is not *G*' is neither true nor false.

I include both the Russellians and the Strawsonians among the orthodox. Both would subscribe to (I) above. The Russellians (1919*b*: 179) would subscribe to (II). The Strawson (1950) of "On

[1] This section includes a revision of Atlas (1977*b*), © 1977, by D. Reidel Publishing Co.

Referring" would probably not, since he commits himself to a "natural" reading of the sentence 'The F is not G' that is the predicate-negation reading. In van Fraassen's (1966, 1971) formalization of Strawson's (1950) view, the distinction between exclusion and choice negation (and between sentence and predicate negation) is made explicitly. Proposition (IV) expresses the Strawsonian view of semantic presupposition in its more precise version, typical of the way formal semanticists and linguists would or should state it.[2] I am therefore opposing views in both Strawsonian and Russellian camps.

A pragmatic attitude toward the Russell–Strawson debate is encouraged by the attitude characteristic of W. V. O. Quine, who, in March 1973 at the conference on "Language, Intentionality, and Translation Theory" at the Univerity of Connecticut, remarked (Quine 1974: 478):

> I didn't suppose there was a continuing dispute about how really to analyze English description. I thought that most people now take the attitude that Russell's theory of description is just fine in giving us an improved regimented language, just as the material conditional is a good scientific improvement over the regular conditional, though in both cases there's no faithful correspondence to ordinary English. And that the Strawson kind of thing probably comes closer to typical ordinary English, although you could find examples, I suppose, against it also.

To this Gilbert Harman replied (Quine 1974):

> But I thought there were still a number of people, Grice for example, who think you can defend the Russell analysis as an account of ordinary English. I think David [Kaplan] holds this. . . . [T]he idea is to explain the kind of *apparent* evidence in favor of [Strawson's] theory as not a matter of the meaning or logical structure of the thing . . . but a matter of orderly presentation of discourse. A certain way of putting normally suggests that certain things are taken for granted but doesn't absolutely guarantee them.

I refer the reader to Grice (1981) for a statement of Grice's views. In the same Open Discussion at the Conference Saul Kripke

[2] For a criticism of Lycan's (1984) account of Strawson, see Atlas (1988*a*). The exclusion negation $-A$ is true if and only if A is not true, for every admissible valuation of the language. The choice negation $\sim A$ is true (false) if and only if A is false (true), for every admissible valuation of the language. In a bivalent language exclusion-negation and choice-negation functions are extensionally identical. In a non-bivalent language, i.e. one with truth-value gaps, they are distinct.

lucidly stated the point of view from which philosophers have continued to be puzzled about Russell's and Strawson's theories (Quine 1974: 479–80):

> Where someone puts this question in a form like 'Is the present king of France bald?', the informant may be puzzled and not say 'No'. But if instead you put it to him very categorically, say first specifying an armament program to make it relevant and then saying 'The present king of France will invade us', the guy is going to say 'No!', right? So what type of response you'll get from the informant will in fact depend on cases. There isn't the clear verdict in favor of Strawson that would be imagined by concentrating on a narrower class of cases. So it seems that if you want a uniform theory you have to explain one or the other class of cases away. Or so it might seem at first blush anyway. And then the evidence seems hard to balance between which class of cases should be explained away. A Strawsonian might say "Oh, the 'No!' to 'The king of France will invade us' is not saying 'That's false' but is a rejection of the statement as having a false presupposition.' Because you sometimes say 'No' when something is neither true nor false. On the other hand, a Russellian might say "You're hesitant to give a verdict when a presupposition isn't fulfilled, even though you know it's false." So either man has some class of cases to explain. Isn't that right? Then the verdict isn't clear, and this is where the puzzle in the continuing philosophical debate comes from.

This, I believe, is a splendid account of a major, if not the major, source of the philosophical stalemate between the two theories (as of 1973 and, I would suppose, for most philosophers at the present as well).

Though I shall not elaborate it here, I do wish to comment briefly on what I take to be the causes of the philosophical stalemate. In the history of physical sciences, when two quite plausible theories have battled each other to a draw the stalemate has typically been overcome when it has become clear (i) that the evidence was indecisive because it was the *wrong sort of evidence* to use in evaluating the competing theories. It was the wrong sort for many reasons, but one reason is typically that the evidence does not reveal (ii) that the stalemate arises from a fundamental *error shared* by what were otherwise competing theories. This, I believe, is exactly the case here. Russell and Strawson were both wrong.[3]

Quine's "assent" and "dissent", 'Yes' and 'No', etc. are the

[3] The common error was diagnosed in Atlas (1974, etc.).

wrong sort of linguistic evidence to use in evaluating Russell's and Strawson's theories. Philosophers have simply mistaken the theoretical role and the methodological importance of such evidence in deciding between logico-linguistic theories. That sort of evidence could never permit a scientifically respectable choice to be made between the two competing theories. The years of philosophical dispute were simply a case of bad linguistic science. Since it happens in physics, chemistry, and biology as well, it is forgivable in philosophy, but none the less it is not admirable. (For a similar, independent argument, see Katz 1987: 179–80.)

What is defensible by good linguistic science, I claim, is my heterodoxy:

(I*) English *sentences* of the forms 'The *F* is *G*' and 'The *F* is not *G*' *have no logical forms* of the familiar sort.

(II*) 'The *F* is not *G*' is not ambiguous. It is sense-unspecified for scope of negation. [*The Atlas–Kempson Thesis*].

(III*) The English word 'not' creates neither a structurally nor a lexically ambiguous sentence. If on the orthodox view the entry in the lexicon for 'not' is such that 'not' is ambiguous, on the heterodox view the entry for 'not' is such that 'not' is univocal.

(IV*) 'The *F* is *G*' does not semantically presuppose that there is a unique *F*. *A reading or semantic representation of a sentence is not a bearer of truth-value; it is not a proposition or a logical form*, though it delimits what propositions can be literally meant by the sentence.

The paradigm cases of scope ambiguity involving negation have led to a mistaken semantic description of negative sentences standardly thought to possess presuppositions. These mistaken descriptions assign to a single surface structure two distinct underlying forms. For example, in many dialects of English the *sentence*:

(1) Everyone didn't show up.

has traditionally been assumed to have two meanings associated with two underlying logical forms, the narrow-scope predicate negation $(\forall x) - Sx$ represented in English by:

(2) No one showed up.

and the wide-scope sentence negation $-(\forall x)Sx$ represented in English by:

(3) Not everyone showed up.

This has traditionally been thought to be a genuine ambiguity accountable for by a difference in the scope of negation. Typically it is observed that the narrow-scope predicate negation (2) entails the wide-scope sentence negation (3), but that the converse is not the case. Since (2) and (3) are not logically equivalent and can differ in truth-value, (2) and (3) are said to differ in sense or meaning. Thus (1) has traditionally been thought to have two distinct senses, i.e. to be ambiguous, and has been assigned two distinct logical forms.

It was Russell's, and is presently the standard, view that the sentence:

(4) The king of France is not wise.

is ambiguous, having two senses and two logical forms, one the narrow-scope predicate negation, usually represented in English by:

(5) The king of France is "non-wise".

and the other the wide-scope sentence negation usually represented in English by:

(6) It is not $\left\{\begin{matrix}\text{true}\\ \text{the case}\end{matrix}\right\}$ that the king of France is wise.

Russell would have said that, if true, the two readings are true for different reasons. When (5) is true it is because a unique, extant king fails to be wise. When (6) is true it is either because there is no king or because there are more than one or because a unique, extant king fails to be wise. Typically it is observed that the narrow-scope predicate negation represented by (5) entails the wide-scope sentence negation represented by (6), but not conversely; that the logical form represented by (5) entails that there exists a king and that he is unique, but the logical form represented by (6) does not entail either. Since the affirmative sentence 'The king of France is wise' entails the existence and uniqueness of a French king, it is usually said that both the affirmative sentence and sentence (4) on its narrow-scope reading represented by (5) *presuppose* the existence and uniqueness of a French king. It is said that the wide-scope reading of sentence (4), allegedly represented by (6), does not have this presupposition.

There are two questions that seem never to have been raised

explicitly prior to Allwood (1972, 1977), Atlas (1974), and Kempson (1975). First, do the English sentences (6), which supposedly represent univocally one reading of (4), each have different "understandings", and, further, "understandings" that allegedly constitute two readings, the one (6) allegedly represents and the one that (5) allegedly represents? In short (Allwood, Atlas), are (4) and (6) paraphrases? And then, secondly (Atlas, Kempson), do the intuitive differences in understandings in (4), and possibly (6), constitute a genuine *ambiguity*, or is the difference a case of *sense-generality* in (4) and in (6)?

I shall first simply report a fact about my own and, I believe, ordinary speakers' idiolects. Just as my linguistic intuitions detect a difference between a presuppositional and non-presuppositional understanding of (4), they detect the very same difference in (6). In these respects sentences (6) and (4) behave like paraphrases in my idiolect. *The standard view that (6) conveys only one understanding is more a logician's prejudice than an empirical linguistic judgement.*[4]

The question that is now before us is whether (4) and (6) are each ambiguous between a non-presuppositional understanding, which on the standard view would be "the" reading of (6), and a factive or presuppositional understanding, which on the standard view would be "the" reading of (5). The answer is that they are not ambiguous. They are sense-general.[5]

In defence of my answer I shall appeal to tests for ambiguity and sense-generality discussed by Zwicky and Sadock (1975). One of the difficulties of the example is that the presuppositional understanding entails the non-presuppositional understanding of (4) and (6). This is true for what Zwicky and Sadock (1975) call 'privative opposites'. For example, 'dog_1' meaning "male canine" and 'dog_2' meaning "canine" are privative opposites with respect to the feature of gender. The same is true for 'mother' and 'parent'. One of Zwicky and Sadock's *tests* is *for* ambiguity of *privative opposites*. If the expression is truly ambiguous it ought to be possible to assert the general case and deny the specific case without contradiction. 'Dog' is taken to be ambiguous, so it ought to be possible to say 'That's a dog, but it isn't a dog' or 'That dog isn't a dog' without anomaly. If such cases were anomalous it

[4] A discussion of my view may be found in Steven E. Boër and William G. Lycan (1976: 48–52) and in Horn (1978*b*, 1985, 1989).

would indicate that the expression was not ambiguous. In our cases we have:

(7) { { *?The king of France is not wise* / *?It's not* { *true* / *the case* } *that the king of France is wise* }, *but* { *he* / *the king of France* } *is (not non-) wise.* }

The second conjunct is the denial of the presuppositional understanding, which on the standard view would be the denial of the reading represented by (5). Since I find these sentences semantically out of bounds, we have our first indication that neither (4) nor (6) is ambiguous.[5]

Another test is by "*semantic differentia*". When a sentence has relatively similar understandings, as I suggest (4) does, or (6) does, but these differ only by one's being sense-specified and the other sense-unspecified for some particular semantic feature, e.g. [+ / − FACT], the feature must be such that the lexicons of natural languages can plausibly fail to use it. A lexical feature for the age of the referent of a third-person personal pronoun might be, unlike gender, an example of a lexical feature that English need not adopt. Certainly in English there is no formal marking in a negative sentence for a presuppositional, or non-presuppositional, understanding of the sentence.

If the difference in understanding were very great it would point to ambiguity rather than sense-generality. But, in contrast to the two understandings of 'They saw her duck', the difference in the understandings of (4), or of (6), is not very great. To see explicitly that this is so, recall Russell's argument. Suppose we could enumerate all the individuals who are wise. We look down the list, and we do not find the king of France. Of course the reason we do not may be either that he is extant but fails to be wise (the standard

[5] Sadock (in a personal communication) has suggested that these sentences might not sound so bad to some ears because lexical negations are usually "poseurs", e.g. 'unhappy' differs from 'not happy'. There is a tendency to read 'non-wise' as 'definitely stupid'. Sadock suggests, as a clearer example, that the following sentence is anomalous:

?The king of France is not wise and it's not even the case that the king of France is wise

which it ought not to be if the negative sentence were ambiguous and took its narrow-scope-of-negation reading in the first clause.

account of the truth of (5)) or that he does not exist (part of the standard account of the truth of (6)). But the similarity in the two cases is significant; we do not find the king of France on the list. This similarity is more compelling than the difference. It is at least sufficiently compelling *to shift the burden of proof* of the ambiguity of (4), or, on my view, of (6), on to those parties who claim it.[6]

Of the two syntactic tests that I shall mention, I shall not employ here *the test of transformational potential*. (I leave that as an exercise for the readers of Zwicky and Sadock (1975). The results will confirm that (4), or (6), is not ambiguous.) But I shall consider J. R. Ross's and George Lakoff's *conjunction test*.

This test is an identity test, one employing a transformation whose application to underlying structure requires identity of sense of constituents. Consider a transformation that yields pro-forms

(8) { { *The king of France is not wise* / *It is not* { *true* / *the case* } *that the king of France is wise* }, *and the same (thing) goes for the Queen of England.* }

that are derived by an identity-of-sense transformation from

(9) { { *The king of France is not wise* / *It is not* { *true* / *the case* } *that the king of France is wise* }, *and* { *the Queen of England is not wise.* / *it is not* { *true* / *the case* } *that the Queen of England is wise.* } }

[6] This argument also confronts the claim of polysemy, ambiguity among closely related meanings. A defender of the ambiguity of (4) and possibly of (6) might well want to argue that sentence and predicate negation are closely related, distinct meanings. I would argue that paradigm cases of polysemy are constituted by meanings related in ways quite different from the way predicate-negation and sentence-negation sentences are logically and semantically related. Consider, for example, 'hit' in 'John hit Brian and the highway divider'. (See ch. 2. n. 6, and the related text.) Consider also 'knocked' in Rogers Albritton's example:

He knocked her over, knocked her up, knocked her out, and then went around knocking her.

Predicate and sentence negations are, in Zwicky and Sadock's (1975) sense, privative opposites. The meanings of polysemous expressions are typically not

If the negations in these sentences are sense-non-specific for "scope", we have four possible understandings of the conjunction: factive/factive, factive/non-factive, non-factive/factive, and non-factive/non-factive. On the ambiguity hypothesis the identity-of-sense condition on the pro-form 'the same (thing)' in (8) should eliminate the second and third crossed understandings.

My idiolect accepts the crossed understanding non-factive/factive:

(10) (*a*) { *The king of France is not wise* / *It is not* { *true* / *the case* } *that the king of France is wise* }, *(since France is not a monarchy), and the same thing goes for the Queen of England (who is a typical Windsor).*

If crossed understandings are acceptable and the sentence is derived by an identity-of-sense transformation, then the instances of constituent type in underlying structure have the same sense if not the same understanding. This means that the constituent is not ambiguous.

The same conclusions hold for adverbials other than 'not'; in these cases the presupposition is on the verb phrase:[7]

(*b*) *John didn't walk slowly (he walked rapidly) and the same goes for James (he stood still).*

(*c*) *John didn't butter the toast with a knife (he used a spoon) and the same goes for James (he stabbed it).*

There is a complexity in applying this test. Zwicky and Sadock (1975) claim that if the constituent is ambiguous and related in sense as genus to specie the identity-of-sense test for ambiguity does not "prove" that the constituent is ambiguous. They consider the example 'dog', which they take to be ambiguous between "male canine" and "canine". These readings are examples of "privative opposites", 'dog' and the like being unusual cases. Nevertheless, this does not mean that 'not' might not be a case of polysemy like 'dog'. So I shall discuss this issue in some detail in what follows.

[7] The presuppositional aspects of these sentences were ignored in the Russellian treatment Donald Davidson (1967*a*) offered of the logical form of action sentences. They were also ignored in the lengthy debates that ensued. Ignoring them was, from the point of view of an adequate semantic theory of natural language, a major error.

privative opposition; one expression is more specific and entails the other more general expression, but not conversely. The question is, is

(11) *I saw a dog.*

ambiguous between these two understandings? To decide the question we apply an identity-of-sense transformation to:

(12) *I saw a dog and Harold saw a dog.*

to yield:

(13) *I saw a dog and so did Harold.*

If (11) were sense-general rather than ambiguous, then so would (12) be. It would have four understandings that would be preserved by the identity-of-sense transformation in (13). Linguistic intuition suggests that (13) has but two of the four understandings. Hence (11) is ambiguous.

Zwicky and Sadock (1975) raise the question of support for the intuitions about (13). How can one show that the crossed understandings:

(14) *I saw a* dog_1 *and Harold saw a* dog_2.
(15) *I saw a* dog_2 *and Harold saw a* dog_1.

are impossible in

(16) *I saw a dog and so did Harold*?

They argue that any context in which a crossed understanding is "appropriate", e.g., I seeing a male canine and Harold seeing a canine, or I seeing a canine and Harold seeing a male canine, is also a context in which the parallel understanding "I see a canine, and Harold sees a canine" is also "appropriate". This is so, on their view, because of the entailment from '*x* is a male canine' to '*x* is a canine'. They argue that since there is no context in which ONLY the crossed understandings are "appropriate" one cannot exhibit the anomaly of the crossed understandings in a way that will convince anyone with different intuitions. He presumably would not find the crossed understandings anomalous.

There is something unconvincing in this argument. Zwicky and Sadock (1975) have used 'appropriate' where it is appropriate to use 'true'. An understanding of a sentence may indeed be true in a context without being an appropriate understanding of a sentence in that context. Even if there are no contexts in which crossed understandings are true and the parallel understanding, with 'dog' as 'canine', is false, the parallel understanding need not be appropriate. The inappropriateness is a function of the sense, not the reference, of 'dog'. For that reason what makes (16) inappropriate may simultaneously make (14) or (15) true.

Contexts may be described in lots of ways. What matters is the particular description chosen, not whether some other description is also true of that context. Like actions, judgements of the anomalousness of a sentence in a reported context, which is reported by a description, need not be preserved for the same context under a different description. Whoever thought they should be?

Finally, there is something wrong in attempting to support intuition in this way. Not only is reliance upon extensionality in characterizing contexts hopeless, but the intuitions do not need, nor can they get, this kind of support. I believe it would be useful to look at this identity-of-sense transformation again with a less puzzling example.

There is nothing wrong with saying:

(17) *I went to the bank*$_1$ *and Harold went to the bank*$_2$.

and meaning:

(18) I went to the currency depository and Harold went to the edge of the river.

But you cannot mean this by saying:

(19) *I went to the bank and so did Harold.*

From that it follows:

(20) Harold did what I did.

which would not be the case above. What counts as satisfying the identity claim for Harold's and my actions is a matter for theory. For the ordinary man in the street no doubt an intensional equivalence between the descriptions of the actions is necessary. Likewise, if one says:

(8)
$$\left\{\begin{array}{l}\left\{\begin{array}{l}\textit{The king of France is not wise}\\ \textit{It is not} \left\{\begin{array}{l}\textit{true}\\ \textit{the case}\end{array}\right\} \textit{that the king of France is wise}\end{array}\right\},\\ \textit{and the same (thing) goes for the Queen of England.}\end{array}\right\}$$

what counts as the same thing is again the issue. If intensional equivalence between descriptions of kinds of states of affairs is again a necessary condition according to the "theory" of our ordinary speaker, we know that *one kind* of state of affairs must be exhibited, but we do not thereby know what counts as states of affairs being of one kind. If the criteria for states of affairs being of a kind are loose enough, we shall accommodate the crossed understandings, as in (10).

The identity-of-sense transformation is the linguist's counterpart to the ordinary man's standards of equivalence. These standards are meant to preserve a practical notion of sameness for application to events and situations in ordinary life, about which we do much of our talking. Identical-seeming descriptions of non-identical circumstances must, *ex hypothesi*, be implicitly non-equivalent and hence "ambiguous". This is the ordinary man's and the ordinary linguist's notion of ambiguity. Putting it somewhat schematically: differences of reference entail differences of sense. Even if this thesis is true, and it is not clear that in general it is—in Chapter 4, Section 2, I argue that it is not—it obviously does not entail that any particular reference corresponds to any particular sense. What reference corresponds to what sense is, in part, a matter of what I have been calling criteria of identity. Whatever they are, they determine what *counts* as a difference in reference, what counts as states of affairs being of one kind.

In contrast to the 'dog' example, where one is assuming that 'dog' is indeed ambiguous (and the identity test tends to support this view), the intuitions in example (8)/(10) support the crossed understanding. If someone does not agree with these intuitions there is no further case to be made for them. This does not mean, however, that the acceptability of crossed understandings for the pro-form (8) produced by the identity-of-sense transformation does not support the hypothesis that the sentence is sense-general rather than ambiguous. Of course it does.[8]

[8] I also criticized Zwicky and Sadock's (1975) discussion of privative opposition in ch. 2, sect. 2. The criticism in the preceding seven paragraphs, originally given in Atlas (1977*b*: 328–30), has recently been supported by similar criticism offered independently by Lawrence D. Roberts. See Roberts (1984), Zwicky and Sadock (1987), and Roberts (1987). Blackburn (1983), failing to see that Atlas (1977*b*: 328–30) was a criticism of Zwicky and Sadock (1975), accuses me of misunderstanding Zwicky and Sadock (1975), misconstruing my divergence of belief from Zwicky and Sadock as a divergence from the meaning of their words. For other discussion

According to these four tests of transformational potential, of conjunction reduction, of semantic differentia, and of anomaly of privative opposites, the difference between presuppositional and non-presuppositional understandings of (4), or of (6), is not a difference in sense. The sentence is not ambiguous; it does not need two logical forms. It does not contain syntactical "scope ambiguities" with respect to 'not'. Nor does it contain a lexically ambiguous 'not', one that is an exclusion negation and another that is a choice negation. These four tests suggest that the correct view is that neither (4) nor (6) is ambiguous; each is sense-general.[9]

But does this matter? I think it does. For a non-bivalent language containing choice/predicate negation van Fraassen defines semantical presupposition as follows: A presupposes B if and only if $A \Vdash_{\mathscr{V}} B$ and $\sim A \Vdash_{\mathscr{V}} B$ where $\sim A$, the choice negation of A, is true (false) if and only if A is false (true) under every admissible valuation $v \in \mathscr{V}$ of the language. Exclusion/sentence negation is not appropriate for defining semantical presupposition. The presupposition B is not entailed by the exclusion negation $-A$. It is choice negation $\sim A$ that preserves presupposition. Thus it is assumed that negation in English is ambiguous between choice and exclusion negations.

The reason is allegedly clear. If 'The king of France isn't wise' can be true when there is no king of France, but 'The king of France is wise' semantically presupposes that there is a king of France, the negative sentence must be ambiguous. The only obvious source of the ambiguity is 'not'.

The linguist Lauri Karttunen (1973*a*: 87 n. 18) comments on the distinction between the different negations:

> This is an important distinction for ordinary language. However, the most natural way to capture it in our framework is to distinguish between two senses of 'not'. As internal negation (choice negation), 'not' is a hole and lets through all of the presuppositions of the sentence it negates. The external 'not' (exclusion negation) is a plug that blocks off all of them.

In another essay Karttunen (1973*b*) remarks that 'John did not have to stop beating his wife' is "ambiguous between the two kinds of negation, internal (choice) and external (exclusion) negation, of

of my view, see John N. Martin (1982*a*, *b*) and comments by Zwicky and Sadock (1984).

[9] For definitions of exclusion and choice negation, see n. 2 above.

which the latter kind voids all presuppositions of the negated clause".

A critic of the notion of semantic presupposition might challenge the presence of a semantic presupposition in a sentence by claiming that in some contexts of utterance the sentence does not have the presupposition. (Semantic presupposition is assumed to be invariant with respect to context.) Richmond Thomason (1973) has stated that in his view the only way to rebut evidence for the contextual dependence of a presupposition is to claim that the sentence is ambiguous. He has written: ". . . negation is semantically ambiguous. Unless this claim is made, I do not believe that the semantic notion of presupposition can be defended at all."[10]

If Thomason is right, and the defensibility of the semantic notion of presupposition requires that negation in English be ambiguous, then if negation is not ambiguous there is no semantic presupposition in English sentences of the form 'The *F* is *G*'. But that simply means that when there is no unique *F* it does not follow that 'The *F* is *G*' is neither true nor false. Defenders of truth-value gaps no longer have a notion of semantic presupposition, or van Fraassen's (1966, 1971) technical surrogate, to appeal to in their defence.

Those who think there is some point to truth-value gaps always ignore Scott's (1976) sceptical arguments against many-valued logic and always mimic arguments of Quine's that purport to show why the distinction between False and Null is a distinction with a difference. Quine (1960: 177) writes:

> Even . . . truth-value gaps can be admitted and coped with, perhaps best by something like a logic of three truth values. But they remain an irksome complication, as complications are that promise no gain in understanding.

Let it not be supposed that these various complexities and complications

[10] The most recent and comprehensive treatment of the theme of context and presupposition is van der Sandt (1988). For relevant discussions of negation, see Allwood (1977), Atlas (1975*a*/*b*, 1977*b*), Kempson (1975), Wilson (1975), Gazdar (1979, 1980), and Horn (1985, 1989). For discussions of my views, see Boër and Lycan (1976), Hollings (1980), Blackburn (1983), Martin (1982*a*, *b*), Zwicky and Sadock (1984), Roberts (1984, 1987), Sadock and Zwicky (1987), Horn (1978*a*/*b*, 1984*a*, 1985, 1989), Seuren (1985), van der Sandt (1988), and Kempson (1988). It is not the purpose of this book to give detailed replies to my critics; I shall do that elsewhere.

issue merely from a pedantic distinction between what is false and what is neither true nor false. Nothing would be gained by pooling these two categories under the head of the false; for they are distinguished, under whatever names, in that the one category contains the negations of all its members while the other contains the negations of none of its members.

The negation of which Quine speaks is obviously choice negation. So far as English is concerned Quine's argument has force if 'not' is assumed to be a choice negation. But the assumption that it is begs the interesting semantic question. This argument should not be mimicked.

I shall discuss truth-value gaps further in Chapter 3, Section 2. In concluding this section I mention briefly a further theoretical consequence of drawing the distinction beween ambiguity and sense-generality. If a negative *sentence* were ambiguous, it should have distinct semantic representations. But if it is not ambiguous *what semantic representation should it have?* For, what is still true about the sentence is that a speaker/hearer can understand it in more than one way. An utterance *The king of France is not wise* can be understood to assert that the king of France is such that he fails to be wise (an understanding that normally "presupposes" that the king exists). It can also be understood to deny the proposition that the king of France is wise (without the existential presupposition). If understanding the *sentence* 'The king of France is wise' involves a pairing of surface and underlying structure with associated semantic representation, what semantic representation is involved? Since the sentence is not ambiguous, the different understandings of an *utterance* of the sentence cannot be explained by appeal to two different semantic representations. *How then are they to be explained?* These questions will be my concern in Chapter 4.

But now that we have reached a very unorthodox conclusion concerning the semantic properties of one of our language's most basic logical words, perhaps it will be comforting to recall a remark of Quine's (1974: 484) that "the English 'not' obviously is not the sign of negation in the logical sense".

But I have never been satisfied by knowing what 'not' is *not* a sign of; I have wanted to know what 'not' *is* a sign of. That is what I have tried to explain here. The result is not what two millennia of logical and philosophical doctrines would have led one to expect. The result is not, I think, obvious.

In the next section I continue the discussion of truth-value gaps, consider some recent ingenious attempts by Formal Semanticists to have Bivalence and the effect of truth-value gaps too, and reconsider the linguistic phenomenology of presuppositional statements in the light of the sense-generality of negative sentences.

2. Truth-Value Gaps and Topic/Comment of Statements[11]

Research in the foundations of mathematics led Frege, Russell, Tarski, and Lukasiewicz to consider the similarities and differences between the language of mathematics and the vernacular. To what extent was a theory of meaning for one also a theory of meaning for the other? Three-valued logic and supervaluational languages have standardly been employed to regiment sentences for which the "presupposition" that their singular terms have referents is false. The adequacy of these semantic accounts can be seriously questioned. The methods of Chomskyan Transformational Generative Grammar and the results of philosophical analysis suggest that the best theory of "presupposition" is pragmatic, not logical, and that the concepts of falsity and negation for a natural language were misunderstood by Aristotle, the Stoic logicians, Frege, DeMorgan, Russell, and Lukasiewicz.

Non-bivalent formal languages, whether their semantics employs partial valuation functions, or valuations whose range consists of three values, or, as more recently in the case of a bivalent language, valuations whose range consists of four "two-dimensional" values (Bergmann (1977, 1981), Herzberger (1973, 1975), Martin (1975)), have been employed (i) to describe intuitions about the application of the ordinary terms 'true' and 'false' to sentences of English, and (ii) to describe intuitions about inferences to one English sentence from another. For example, it has been said that the sentence 'If it is fine tomorrow, I shall go out for a walk' is true if it is fine and I walk out, false if it is fine and I do not walk out, but when it rains and I remain at home, it expresses no proposition and so is neither true nor false. The question of truth does not arise (Waismann 1945–46). It has also been claimed that a speaker who currently uses a sentence of the

[11] This section includes a revised version of Atlas (1981), © 1981, by The Institute of Electrical and Electronics Engineers, Inc.

form 'The *F* is *G*' to make a statement, which is intended to be a *semantically referential, singular assertion*, viz. that the individual denoted by the definite description 'the *F*' has the attribute indicated by the predicate symbol '*G*', simply fails to fulfil the intention when the description is improper. The statement cannot count as a semantically referential, singular assertion; so it does not count as true or as false. The question of the assertion's truth does not arise, because the question of the statement's counting as that sort of assertion has arisen and has been settled, negatively, by the failure of the singular term to denote an individual (Atlas (1975*b*), Strawson (1950, 1952, 1964)).

Other classes of cases for which failure of truth-value has been regarded as plausible include examples like *Mart regrets having misled Laura* and *John managed to fool Rebecca*. The former statement, employing the "factive" predicate 'regret', is said to be true or false only if *Mart has misled Laura* is true (Kiparsky and Kiparsky (1970), Karttunen (1971*a*, *b*)). The latter statement, employing the "implicative" predicate 'manage', is said to be true only if *John fooled Rebecca* is true, and false only if *John did not fool Rebecca* is true. But in addition 'manage' conveys the "presupposition" that the agent overcomes an obstacle in performing his action. It has been claimed that when this "presupposition" does not obtain the statement *John managed to fool Rebecca* is neither true nor false.

The second sort of data consists in the inferences it is natural to draw from pairs of affirmative and main-verb-negated sentences. When used assertorically both members of the pair:

(21) (*a*) *The queen of England is fat.*
     (*b*) *The queen of England isn't fat.*

seem to convey the "presupposition" that there is a queen of England. An assertion made by a use of the sentence 'The queen of England isn't fat', if it counts as a semantically referential, singular assertion of the kind described above, is true if and only if 'non-fat' applies to the denotation of the singular term 'the queen of England'. Thus the semantically referential, singular assertion made by a use of the sentence 'The queen of England is fat' is false if and only if the analogous, contrary assertion made by a use of the main-verb-negated sentence 'The queen of England isn't fat' is true. The question of the truth of the latter statement does not arise unless the definite description is proper, i.e. unless a semantically referential, singular assertion has been made. Both sentences have the same "presupposition", in the sense that the definite description must be proper if the sentences are used to make the kind of assertion in question. Only if such a use of the sentences is possible can truth-values be assigned to what is asserted.

Both (22*a*) and (22*b*) convey the "presupposition" (22*c*):

(22) (*a*) *Mart regrets having misled Laura.*
(*b*) *Mart doesn't regret having misled Laura.*
(*c*) *Mart misled Laura.*

If the inference from (22*a*) or (22*b*) to (22*c*) is due to the meanings of the sentences, an adequate semantic representation of the sentences should account for the inference.

The semantic notion of presupposition is characterized as a relation between sentences:

(23) *A* semantically presupposes *B* if and only if
A is true, or false, only if *B* is true.

In a non-bivalent formal language containing at least a choice negation $\sim A$ for each sentence *A* such that $\sim A$ is true (false) if and only if *A* is false (true) under every admissible valuation $v \in \mathscr{V}$ of the language, it has been convenient to characterize semantic presupposition by:

(24) *A* semantically presupposes *B* if and only if
(i) $A \Vdash_{\mathscr{V}} B$
(ii) $\sim A \Vdash_{\mathscr{V}} B$.

This view of presupposition, well known to formal semanticists, accounts for the inferences by a kind of non-classical entailment relation and for the truth-value gap in negative sentences by the difference between choice and exclusion negation (van Fraassen 1966, 1969, 1971). The view diverges from the Strawsonian view that I have outlined. Strawson distinguishes, in effect, among sentences, statements, and assertions. The presupposition is a necessary condition for a particular kind of use of a sentence, viz. its statement-making use, to be a particular kind of assertion. The truth of the presupposition is necessary for the sentence to be able to play a particular linguistic role in our discourse. More specifically, there is a distinction to be drawn between the use of a main-verb-negated English sentence to make a semantically referential, singular assertion and the identification of the truth-conditions of a main-verb-negated English sentence with those of a choice-negation logical form from an appropriate formal language. Formal semanticists have ignored this distinction. They have erroneously viewed their semantic version as an explication of Strawson's view (e.g. van Fraassen 1971).

The difficulties of the identification are immediately evident. The statement:

(25) *Mart doesn't regret having misled Laura, because, in fact, he didn't mislead her.*

is acceptable and consistent, which it would not be were there a choice negation in the first clause. The semantic solution normally is to distinguish (with Aristotle, the Stoic logicians, DeMorgan, and Russell) choice from exclusion negation: to give the sentence different logical forms. The two negations are distinguished by their different truth-tables or, as in Russell's case, by making a syntactic scope distinction in the employment of a single operator on (open or closed) sentences. These strategies amount to positing a lexical or syntactic ambiguity in the negative English sentence: the extensional meanings of the sentence are represented by different well-formed formulae in the chosen formal language. Before 1974 philosophers or logicians would have entertained few doubts that English 'not' was adequately represented, at least in its employment in simple, subject-predicate sentences, by the standard negation operators in classical logic, intuitionistic logic, or in the three valued logics of Lukasiewicz (1967), Bochvar (1939), and Kleene (1938). (Exceptions include F. H. Bradley, Friedrich Waismann, and Peter Strawson.) Nevertheless:

(i) 'not' is not ambiguous between choice and exclusion negation;
(ii) 'not' is sense-unspecified between choice and exclusion negation.
(iii) So, exclusion negation is not a sense of 'not', and choice negation is not a sense of 'not'.
(iv) Hence, 'not' is not semantically identical to either exclusion or choice negation.

Observation (i) shows that semantic theories of presupposition that assume two senses of negation are incorrect semantic descriptions of natural language. Observation (iv) shows that the identification of 'not' in English with one of the logical negations would also be incorrect. The sense of 'not' is not adequately represented by a "normal" negation: "internal" (in the sense of Lukasiewicz, Bochvar, Kleene), "external", or "intuitionist" (Lukasiewicz).[12]

Various inadequacies in the treatment of linguistic data have cast doubt on the semantic conception of presupposition (see Gazdar (1979), Kempson (1975), Wilson (1975)), but the observations just mentioned refute that conception as an account of presupposition, and so challenge the view that many-valued logic provides a linguistically adequate account of the logic of affirmative and negative sentences containing non-denoting singular terms or other presupposition-inducing constituents.

It is little short of extraordinary that discussions of presupposition and truth-value gaps have largely ignored features of the uses of presuppositional sentences that were pointed out by Paul Grice, Geoffrey Warnock, and Peter Strawson in the mid-1950s, and later emphasized by Strawson in a well-known essay (Strawson 1964). These features illumine the nature of presuppositional sentences and help show why the semantic account is defective.

The words 'Rick ran away' may be used in statement-making, as contrasted with their use in questioning: 'Rick ran away?', but what assertion is made depends not just on the string of words but also on the way in which they are uttered, e.g. on stress and intonation:

(26) (*a*) *Rick ran away.*
 (*b*) RICK *ran away.*

In making an assertion a speaker is "commenting" upon a "topic". The grammatical subject of the sentence is 'Rick'. In an assertion expressed by (26*a*) the thematic subject is 'Rick'. The topic of that assertion is Rick. The comment is that he ran away. In an assertion expressed by (26*b*), the thematic subject is 'ran away'. The topic is running away. The comment is that Rick did it. (See Lyons 1977: 500–11.)

The same contrast in thematic structure is marked by the difference in syntax between:

(27) (*a*) *Rick ran away.*

[12] See Rescher (1969).

(*b*) *It was Rick who ran away.*

In statements made by uttering the cleft sentence (27*b*), the topic is running away, and the comment is that Rick did it. (Notice that the usual negation of (27*b*) 'It was not Rick who ran away' is not what is meant by the usual negation of (27*a*) 'Rick did not run away'. (27*b*) "presupposes" that someone ran away; (27*a*) does not "presuppose" this.)

If we consider the sentence 'The king of France is bald' in the light of the various assertions it can be used to make, we note that the thematically unmarked statement:

(28) *The king of France is bald.*

the contrastively stressed, thematically marked, statement:

(29) THE KING OF FRANCE *is bald.*

and the thematically marked statement:

(30) *It is the king of France who is bald.*

have as thematic subjects 'the king of France', 'bald', and 'bald' respectively. Expressions containing new information are stressed; expressions containing old information are unstressed. The intuitions that caused Strawson to claim that the unmarked statement is not an assertion and so neither true nor false also suggests that the assertions made in uttering (29) and (30) are both false rather than neither true nor false. As Strawson (1954, 1964) observed, the difference evidently lies in the different thematic structures of the statements. In the case in which it is plausible to say that the statement is neither true nor false, the statement has no topic. But the sentences all have grammatical subjects, and the assertions made by stating (29) and (30) are "about", i.e. have as topic, the bald.

Consequently, the occurrence of a non-denoting singular term in a sentence does not imply that the sentence is semantically anomalous. Some sentences containing non-denoting singular terms are nevertheless used to make assertions that have topics about which the comments are either true or false. Both the sentence and the illocutionary act are in order. By contrast statements that lack topics are not assertions. Such statements are illocutionally anomalous (see Lyons 1977: 601–2).

The distinction I wish to maintain between the semantic properties of sentences and the illocutional properties of statements is not just Strawsonian dogma. As I have argued elsewhere (Atlas and Levinson 1981), English sentences having the forms:

(31) (*a*) If it was *x* that *F*, then *xF*s.
(*b*) If it wasn't *x* that *F*, then *x* doesn't *F*.

are valid conditional schemata. When the substituend for '*x*' is a non-denoting singular term, as in

(32) (*a*) *If it was the king of France that kissed Mary, then the king of France kissed Mary.*
(*b*) *If it wasn't the king of France that kissed Mary, then the king of France didn't kiss Mary.*

the familiar, many-valued matrices for the conditional yield unacceptable results. If *T* is the designated value, the antecedent of (32*b*) is intuitively *T*, and the consequent of (32*b*) is intuitively *N*, the conditional receives an undesignated value whether the trivalent logic is Lukasiewicz's, Bochvar's, or Kleene's. But a valid conditional should not receive an undesignated value.

Bergmann's (1977, 1981) recent four-valued, "two-dimensional" treatment, where on her view the antecedent and consequent of (32*b*) would be false and "semantically anomalous" because of the non-denoting singular terms, makes the conditional true and "semantically anomalous". This gives a better result for the truth-value of the conditional than the trivalent matrices do, but it misdescribes the linguistic situation. An unmarked statement of the antecedent of (32*b*) is true and acceptable. On Strawson's view an unmarked statement of the consequent would be illocutionally unacceptable and therefore neither true nor false. In the case of sentences containing non-denoting singular terms, *N* is an undesignated value. It would be correct to view *N* as a kind of falsity (see Dummett 1959). As Bergmann's valuations are bivalent on each dimension, the truth dimension and the anomaly dimension, it is reasonable to suggest that on something close to her view the consequent of (32*b*) would be false and "semantically anomalous". Bergmann's matrix for the conditional would then make the conditional false and "semantically anomalous". But on the basis of both semantic analysis and strong linguistic intuitions, the conditional as a whole is grammatical, and its statement is illocutionally acceptable and intuitively true.

So I conclude that neither a three- nor a four-valued formal language seems to provide quite the linguistic model we need, even where, as in Bergmann's account, a distinction is made between bivalent valuations of truth-value and bivalent valuations of anomaly-value.

It is surprising that Bergmann, having separated truth-value from semantic anomaly (i.e. reference from sense), introduces as the formal counterpart of English main-verb negation an internal-negation connective that recombines valuations of truth-value and anomaly-value. Roughly, on her view, the internal negation of *A* is true if and only if the external negation of *A* is true and *A* is "non-anomalous". This analysis of the sense of internal negation is at the very heart of Bergmann's theory (Bergmann 1977: 65–6; 1981: 33, 38).

The analysis has the effect of allowing the double negation principle for internal negation:

(33) $A \;\not\Vdash\; -\,-\,A$

to fail, since $-\,-\,A$ is equivalent to *A and 'A' is non-anomalous*. The failure of Double Negation for Bergmann's internal negation should raise a modest question of principle: is this negation? But there is a second difficulty, this time a question of application.

On Bergmann's view the sentence:

(34) The king of France is bald or the President of the United States is a Republican.

is true (in 1988) but contains a non-denoting singular term, and so is "anomalous". Its double (internal) negation is false. Bergmann (1981: 33) would think of the double negation of this disjunction as saying that the king of France and the President of the United States are such that the former is not-not-bald or the latter not-not-Republican. But then why wouldn't one have understood the disjunction itself as saying that the king of France and the President of the United States are such that the former is bald or the latter a Republican? In other words, why is the logical form of the disjunction $A(t) \vee B(t')$ rather than $\hat{u}\hat{w}(A(u) \vee B(w))(t,t')$ if $-\,-\,(A(t) \vee B(t'))$ is understood as $-\,-\,\hat{u}\hat{w}(A(u) \vee B(w))(t,t')$ rather than as $-\,-\,A(t) \vee -\,-\,B(t')$? Does the application of double negation suddenly transform the logical structure of the disjunctive sentence? If not then, contrary to Bergmann's claim, the disjunctive sentence (34) would be false rather than true. The sentence would not be a counter-example to (33), but it would then fail to serve Bergmann's purposes.

The disjunction is important to Bergmann's (1981: 34–5) position because on it rests the alleged advantage of the four-valued, "two-dimensional" approach over Bochvar's or Kleene's three-valued logic. Bochvar's disjunction would be *Null*, and Kleene's disjunction would be *True* for sentence (34). If Bochvar takes *Null* to mean *anomalous*, and Kleene takes it to mean *non-true*, it is understandable that (34) would be thought by Bochvar to be *N* and by Kleene to be *T*. Bergmann believes that she can avoid this dispute over which logic to choose, and so having to choose between saying that the sentence is anomalous and that the sentence is true, by making questions of truth-value independent of questions of anomaly and claiming that the sentence is BOTH "anomalous" and true.

But, to repeat, if the sentence is true, but "anomalous", its Bergmann double (internal) negation is false (and "anomalous"). The failure of Double Negation raises some question as to the acceptability of Bergmann's definition of internal negation (see Rescher 1969: 129). On the other hand, as the sentence and its double negation are "anomalous" because of the SAME occurrences of non-denoting singular terms, and the "anomaly" makes the double negation false, by parity of reasoning the semantic representation or logical form of the affirmative sentence should make it false too. Thus the claim that the disjunctive sentence (34) is BOTH true and "anomalous" implies that Double Negation fails to preserve truth and that the double negation operator magically alters the logical form of the disjunctive sentence.

Finally, Bergmann's matrices imply that the conditional

(35) If there is a king of France, then he's bald.

is also both true and "anomalous". Intuitively there is no linguistic anomaly in this sentence at all. The sentence as a whole does not convey a false "presupposition" that there is a king of France.

No systematic advantage seems to accrue to describing these examples as BOTH true and "anomalous" (because containing non-denoting singular terms). Improper descriptions do not always create truth-value gaps (e.g. 'It's the king of France that is bald' is false. See Atlas and Levinson (1981).) The anomaly, if any, is illocutional; it depends on the thematic structure of the assertion made by the use of the sentence.

We have seen that there is an illocutional difference between the unmarked statement-making use of the sentence:

(36) The king of France is bald.

which fails to be a semantically referential, singular assertion, and the marked statements:

(37) (*a*) *THE KING OF FRANCE is bald.*
(*b*) *It's the king of France that is bald.*

which are false statements about the bald. Likewise, there is an illocutional difference between one unmarked statement-making use of the sentence:

(38) The king of France isn't bald.

which fails to be a semantically referential, singular assertion, and the similar (but not identical) marked statements:

(39) (*a*) *THE KING OF FRANCE isn't bald.*
(*b*) *It isn't the king of France that is bald,*

which are true assertions about the bald. None of these is the same statement as the denial:

(40) *The king of France is NOT bald,*

which is about the proposition that the king of France is bald, and is a correct dissent from it. Furthermore, none of these statements is the same as an unmarked use of the sentence:

(41) The king of France isn't bald.

to make a true, non-singular assertion that is the external negation of the proposition that the king of France is bald. All of these statements employ the same sentence, whose meaning is unspecified for, rather than ambiguous between, predicate and sentence negation or between object-language negation and metalanguage denial. It is precisely because the meaning of the 'not'-sentence is sense-general that the same string of words, like a two-dimensional drawing that can be differently interpreted as a projection of differently oriented three-dimensional objects, can be differently interpreted as making different statements. But it does not follow that these different interpretations constitute an ambiguity in the sentence.

Like the Necker-cube drawing, where the two dimensional representation radically underspecifies from what perspective we shall "see" a cube as represented, the literal meaning of the sentence radically underspecifies what statement it is used to make. A truly "ambiguous" figure, like Jastrow's duck-rabbit drawing and unlike Necker's cube drawing, does not radically underspecify its different interpretations; it specifies radically different interpretations.

The absence in the general, lexically and syntactically unambiguous sentence of syntactical and semantical markers that might specify different interpretations *explains* why contrastive stress and intonation play prominent roles in uses of the negative sentence. Our interpretations depend not on the syntactic and semantic but on the possible phonetic representations of the sentence uttered. The range of interpretation includes statements that are choice negations, exclusion negations, and metalinguistic predications. 'Not'-sentences are semantically less specified, and theoretically more complex, than the two-thousand-year tradition in logical theory has heretofore recognized.

3. Topic/Comment and Negative Existence Statements[13]

From Bertrand Russell (1903) to David Pears (1967), and beyond, philosophers have thought that *John Walter exists* was about John Walter, even though the same intuition that *John Walter does not exist* is about John Walter has led them to extravagant metaphysical claims, e.g. Pegasus and the golden mountain have being (Russell 1903). In deflating the ontology, that same intuition has led them to equally extravagant semantic claims, e.g. (i) that *John Walter does not exist* is meaningless (Russell 1919*a*/56: 252) while *The author of "Dog Soldiers" does not exist* is meaningful, (ii) that neither *The author of "Dog Soldiers" is talented* nor *The author of "Dog Soldiers" does not exist* is ABOUT the author of "Dog Soldiers" (and for exactly the same logical reason; see Russell 1919*a*/56: 250), and (iii) that a proper name 'John' is "really" (i.e. in logical form) a predicate 'Johnizes' (Quine 1960) or is "really" (i.e. in a theory of truth for English) a predicate cum demonstrative (Burge 1973).

A famous attempt to control Russell's linguistic extravagance, proposing that statements of

(42) *The author of "Dog Soldiers"* $\left\{\begin{matrix} \textit{is} \\ \textit{is not} \end{matrix}\right\}$ *talented.*

[13] This section includes a revised version of Atlas (1988*b*), © 1988, by D. Reidel Publishing Co.

presuppose the existence of the author of "Dog Soldiers" (Strawson 1950, 1964), led to further extravagances. For example, the statement:

(43) *The great American novel does not exist.*

apparently presupposes that the great American novel does exist, so that if the statement is true, then it is neither true nor false; so, if it is true, it is not true; so it is necessary that it is not true. However much American optimism would be buoyed by a logical refutation of negative European literary assessment, the problem of American literature is not to be solved by logic alone.

Similarly,

(44) *Pegasus does not exist.*

is true-or-false only if Pegasus exists, and so is neither true nor false rather than exhibiting the truth we know and love. There are philosophers who could convince themselves that *Pegasus does not exist* is neither true nor false because 'Pegasus' has no "superdenotation".

Russell (1919*a*/56: 242) and Strawson (1950, 1952) would be in agreement here, with the proviso that one were treating 'Pegasus', like 'Romulus', as an *apparent* name and not a truncated definite description. Since there is no winged horse, 'Pegasus' is a vacuous name, and on Russell's view a meaningless name, and so not a name at all! Thus 'Pegasus does not exist', where 'Pegasus' is not a truncated description, does not express a significant singular proposition in Russell's view, since it is in fact missing the constituent corresponding to the grammatical subject of the sentence. So no true or false proposition has been expressed. On Strawson's view, since the presuppposition that Pegasus exists is false, no true-or-false statement is made. The oddity that neither 'Pegasus exists' nor 'Pegasus does not exist' could be significant, i.e. true or false, unless Pegasus exists pushes Russell to the conclusion that Pegasus is not a constituent of those propositions. Then Russell wonders what one does mean by 'Pegasus does not exist', given that he has already decided that existence is a second-order property. So he resolves Pegasus into properties (or 'Pegasus' into predicates) in his theory of definite descriptions.

If, fancifully, we give 'Pegasus' a "superdenotation" but deny it a denotation, we might evade the semantic problem that the Strawsonian repair of Russell's linguistic artificiality has created:

semantically motivated neo-Meinongianism. Russell (1903) (and recently T. Parsons 1980) had a semantic motive for positing being. That created an antithetical metaphysical motive for contextually defining singular terms. That led to an antithetical reassertion of the logically primitive character of referring terms. That becomes, in my fancy (and in others' serious philosophy), a semantic motive for positing being ("superdenotations", and the like). This is eighty-five years of philosophical progress! So I want to go back to the very beginning: Is *John Walter exists* about John Walter?

I shall appeal to Gundel's (1977) discussion of statement topic/comment to show that non-emphatic, informative statements of the sentence 'John Walter exists', and of 'John Walter does not exist', are not about John Walter, because in those statements tokens of 'John Walter' are not topic-designating Noun Phrase tokens. The topic is: what exists, or, as Quine would say, everything.

I shall show that this thesis concerning the topic-designating Noun Phrases of affirmative and negative existence statements permits a complete solution to the notorious anomaly, sketched above, in Strawson's account of the presuppositions of existence statements.

This thesis also provides a linguistic *explanation* for the appeal, such as it is, of Russell's theory of definite descriptions to our linguistic intuitions concerning the paraphrase of the *The F is G* by *There is an F, and at most one F, that is G*.

It is my view that philosophical argument about affirmative and negative existence statements has, from Russell (1903) to Parsons (1980) and the present, derived entirely from a linguistic mistake.

A. *Quine's Version of Russell's Problem and Strawson's Reformulation*

For Russell the meaningfulness of negative existential statements necessitated the positing of non-being:

(PB) What does not exist must be, or it would be meaningless to deny its existence.

Quine called the doctrine *Plato's Beard* and stated it as follows: "Nonbeing must in some sense be, otherwise what is it that there is not?" In particular, if "Pegasus were not . . . we should not be talking ABOUT [*my emphasis*] anything when we use the word; therefore it would be nonsense to say even that Pegasus is not. Thinking to show thus that the denial of Pegasus cannot be coherently maintained, [one] concludes that Pegasus is" (Quine 1980: 2). Schematically,

(1) (*a*) If N were not, we should not be talking ABOUT anything when we use the name 'N'.
(*b*) So, the name 'N' would be meaningless.
(*c*) So, 'N is not' would be nonsense.

(2) (*a*) In fact, 'N is not' is not nonsense.
(*b*) So, N is.[14]

We can then proceed to use Russell's (1905) theory to resolve the (1903) puzzle of Plato's Beard. Quine formulates the underlying general principle as follows (Quine 1980: 8–9):

(M_R) The meaningfulness of a statement containing a singular term presupposes [the existence of or the being of] an entity named by the term.

The theory of definite descriptions provides Russell's Razor (Quine 1980: 7):

(RR) When a statement of being or nonbeing is analyzed by Russell's theory of descriptions, it ceases to contain any expression which even purports to name the alleged entity whose being is in question, so that the meaningfulness of the statement no longer can be thought to presuppose that there be such an entity.

Russell's puzzle of non-being in the Quinean form:

(NB) 'Pegasus is not' is nonsense because 'Pegasus' is meaningless.

is understood by Quine to rest upon Russell's conflation in (M_R) of meaning with naming. This, I believe, is neither a historically accurate criticism of Russell nor a solution to the semantic problem of explicit negative existentials. It is not a solution

[14] Quine allows proper nouns as substituends for 'N'. Russell thought that *N exists*, where 'N' is an individual constant, was logically meaningless. Russell's solution to the puzzle of negative existence statements in his theory of definite descriptions does not apply directly to *N exists* where 'N' is a proper name. Quine adopts the Russellian solution of converting the name into a definite description. If there is no description readily available because our notion of N is obscure or our use of 'N' is basic in our language, Quine takes as basic the predicate *being N*. The description employed is then 'the thing that N-izes'.

because the problem can be reformulated in only slightly different terms.

Suppose a philosopher, say McZ, replies to Quine that he never made that error. He never confused meaning with naming nor the meaningfulness of a statement with its being true or false. But he has held the view, which I shall call 'The Strawson Point', that when one asserts *The king of France is bald* one has not made a true or false, *semantically referential, singular assertion*, because there is no king of France (see Chapter 3, Section 2).[15] The question of the truth or falsity of a singular assertion just does not arise because with failure of reference no singular assertion has been made. Following Quine's lead, McZ generalizes this observation in the following:

(M_S) 'Pegasus is not' is neither true nor false because 'Pegasus' has no referent.

Russell's puzzle of non-being, in Quine's version, was that "Pegasus cannot be said (meaningfully) not to be without presupposing in some sense Pegasus is" (Quine 1980: 8); McZ's puzzle of non-being is that Pegasus cannot be said (truly-or-falsely) not to be without presupposing that in some sense Pegasus is. Russell's puzzle originated in the argument that if "Pegasus were not . . . we should not be talking ABOUT anything when we use the word; therefore it would be nonsense to say even that Pegasus is not" Quine 1980: 2). McZ's puzzle originates in McZ's argument that if Pegasus were not, we should not be talking ABOUT anything when we use the word; therefore it would be neither true nor false to say even that Pegasus is not.

Quine had argued, independently of Russell's theory of definite descriptions, that the error in Russell's puzzle of non-being could be exposed by a careful attention to the distinction between meaning and naming. This distinction could be used to separate the elusive error concerning meaning in the doctrine of Plato's Beard from issues of existence. But this distinction will not avert the difficulty that McZ has posed. McZ's puzzle of non-being seems just as puzzling as Russell's did. Are we now forced back to the "technological solution", viz. Russell's theory of descriptions,

[15] Unpublished manuscripts and letters of Russell of the summer of 1904 show that at one point Russell himself was such a McZ. For the need to distinguish between sentences, statements, and kinds of statements (called 'assertions' in Atlas (1981)), see Strawson (1964), Atlas (1981), and ch. 3, sect. 2.

and its concomitant analysis of existence? Or can we hope to find a deeper solution to McZ's puzzle through a critical examination of (M_S)?

If I assert 'The king of France is bald', and there is no king of France, contrary to the presupposition, Strawson's Point is that I fail to make a true or a false (attributive) singular assertion. My act of making such an assertion has "misfired"; if I was trying to do that, I failed. On the other hand, if I assert the cleft sentence 'It is the king of France who is bald', and there is no king of France, I do succeed in making a false statement, and hence a statement that is either true or false. If so, evidently the statement does not presuppose, in Strawson's sense, that there is a king of France.

If the reader checks his intuitions he will, I think, agree that what I presuppose in asserting 'It is the king of France who is bald' is that someone is bald. Note, for example, that when I assert 'It is not the king of France who is bald' one may still infer that someone is bald.

The question naturally arises whether a speaker also "presupposes" in asserting 'It is the king of France who is bald' that there is a king of France. But the question whether *the speaker* "presupposes" this is a different question from the one I am asking, which is the question whether *the statement* presupposes this. The question about the statement is about certain linguistic conventions in the use of the cleft sentence to make various statements, which can be distinguished by stress and intonation.

If someone asserts, with contrastive stress, *It is the king of France who is* BALD, I would normally understand *the speaker* to presume that there is a king of France, and if he does not, he is misleading me, or he does not know how to speak English. The speaker may also felicitously assert *It is the Prince of Denmark who is* BALD, meaning to say something about the Shakespearean character Hamlet. Characters in narrative fiction are named or described, and in English we do not use linguistic devices, e.g. a fictional-character morpheme or a hypothetical-entity morpheme (for early uses of 'neutrino' or uses of 'black hole') to indicate the ontic commitment of our uses of names or descriptions. But that is why it is the responsibility of the speaker to mark his discourse as fictional or hypothetical.

In any case *It is the king of France who is* BALD is not the assertion about which I asked my original question. My question was about an unemphatic assertion *It is the king of France who is bald* and asked of it what any English speaker would understand to be a condition on the successful making of such an assertion. Even if some speaker who asserts this presumes that 'the king of France' denotes the king of France, the lack of truth of that presumption does not cause the assertoric speech act to "misfire", i.e. to result in no true or false assertion being made, and I know that because I know speech-act conventions governing this use of the cleft sentence. I know that if the presumption is not true, the assertion is linguistically acceptable but false.

Thus I want to distinguish what in a particular context a particular speaker may presume from what in any context a use of a sentence to make a particular type of assertion may presuppose, as a condition of the successful making of an assertion. It is a condition of the successful making of an attributive, singular assertion *The F is G*, one that purports to be about the denotation of 'the *F*' and asserts of it that it is *G*, that the denotation of 'the *F*' exist. It is not a condition on the *linguistic acceptability* of the statement *It is the king of France who is bald* that the king of France exists.

The reason for this is that in a statement *It is the king of France who is bald* the expression 'the king of France' is not the topical subject; so there is no call for such an entity to be the topic about which one is commenting in making the assertion. The statement would presuppose that 'the king of France' is used to refer to an entity only if the noun phrase 'the king of France' is a topic-designating expression in the statement. So the statement does not presuppose that there is a king of France. In the Introduction I mentioned my concern to distinguish levels of theory, that of sentence grammar, that of conventions of speech acts, and that of context-situated speaker's inference. The distinctions I am drawing here are a case in point.

Note well: if *e* is a topic-designating expression in a statement (not, please also note, a sentence), it does not follow that there is a designatum *d* such that *e* designates *d*, any more than if *p* is a centaur-depicting picture, it follows that there is a centaur such that *p* depicts it. But *d* is a topic of a statement only if a topic-designating expression *e* denotes or is used to refer to *d*.

I wish to distinguish among the denotation by an expression, the presupposition by a statement, the referring by a speaker, and the presuming by a speaker. In statements of both (45*a*) and (45*b*):

(45) (*a*) *The queen of England raises the best race horses.*
(*b*) *It is the queen of England who raises the best race horses.*

the singular term 'the queen of England' may be used by a speaker to refer to the queen of England. And, in fact, in both statements 'the queen of England' has a denotation, viz. the queen of England. But it is only an attributive singular assertion of (45*a*) that presupposes that the singular term is used (attributively) to refer to the queen of England. Even in the case of a non-vacuous term, like 'the queen of England', there is a categorical distinction among a statement's presupposing, a speaker's presuming, an expression's denoting, and a speaker's referring. One should not confuse a speaker's use of a singular term to refer with a particular type of statement's presupposing that the singular term is being used to refer.

McZ's basic argument in the case of statement-type (45*a*) would have been that if the queen of England were not, one should not be talking ABOUT anyone when he uses the term 'the queen of England'; therefore, it would be neither true nor false to say that the queen of England raises the best race horses. In general, McZ would hold to the principle:

(M_S) The truth-or-falsity of a statement containing a singular term presupposes [the existence of] an entity referred to by a use of the term.

Nevertheless, McZ's, and also Frege's, principle (M_S) is refuted by the example of statement-type (45*b*). When a speaker uses the singular term in making this statement, even if there is no queen of England, he does not fail to be talking ABOUT anything. In this statement he is not talking ABOUT the queen of England; she is not the topic of the statement. So this statement does not presuppose that there is a queen of England. Consequently, it does not follow that it would be neither true nor false to say that it is the queen of England who raises the best race horses, even if the queen of England were not. Indeed, were there no queen of England it would be false to say that it is the queen of England who raises the best race horses.[16]

[16] The reader will find an account of the pragmatics and logical semantics of cleft sentences in Atlas and Levinson (1981). For agreement with these intuitions see James McCawley (1981: 241).

McZ's puzzle of non-being:

(NB_S) 'Pegasus is not' is neither true nor false because 'Pegasus' has no referent.

rests upon McZ's conflation in (M_S) of presupposing with referring. The former essentially involves statement-ABOUTNESS; the latter does not. Since ABOUTNESS has been used without analysis by Russell and Quine, it now deserves some attention. I propose that an understanding of it will dissolve the philosophical, semantic problem of existentials.

B. Russell's Appeal to Aboutness

We have just seen that a standard diagnosis of the problem of negative existentials, viz. Quine's, fails to account for an elementary Strawsonian variant of the problem. The fundamental mistake underlying the problem is neither that "existence is a (first-order) predicate" nor that "meaning is naming". It is a linguistic mistake conceptually prior to mistakes in the logical syntax of 'exists' or to mistakes in the meaning and reference of singular terms. I shall show that it is a linguistic mistake as to the "thematic" content of existential statements, viz. a mistake as to what existential statements are ABOUT, or in linguistic terms, as to the "topics" of existential statements.[17]

Sir Karl Popper, Hilary Putnam, and Nelson Goodman offered explications of ABOUTNESS (see Atlas and Levinson 1981), but the notion has not been taken sufficiently seriously by philosophers, confused as they are about the difference between a *statement*'s being *about* something and a *singular term*'s being used to *refer* to something, and intimidated as they are by the alleged indeterminateness of the notion of ABOUTNESS. The former circumstances has made a discussion of the notion seem superfluous, subsumable under a discussion of reference.[18] The latter circumstance has

[17] It is conceptually prior because, like others in the Fregean tradition, I take statements/sentences as fundamental.

[18] The conflation of statement-aboutness with term-reference, and the subsumption of the former under the latter, is just a manifestation of the anti-Fregean commitments of philosophers who otherwise take themselves to understand and espouse Fregean principles, particularly the priority of the sentence. Unlike many, Donald Davidson has been admirably clear on this priority and consistent in his theoretical stance. See his "Reality without Reference" in Davidson (1984: 215–25).

made discussion of the notion seem suspect, subsumable under no discussion rigorous enough to merit serious attention. In my view the former circumstance is the product of an unfortunate linguistic mistake; the latter is the product of an intellectually stultifying fastidiousness.

As I discussed in Atlas and Levinson (1981: 42–3), Putnam (1958) explicated a notion of aboutness for universally quantified conditionals. His basic intuition was that *All Fs are Gs* is about *F*s. He also required that logically equivalent sentences be about the same things, so that *All Fs are Gs*, which is equivalent to *All non-Gs are non-Fs*, is also about the non-*G*s. Since *Fa* is logically equivalent to $(\forall x)(x = a \rightarrow Fx)$, *Fa* is about $\{a\}$. And $- Fa$, which is equivalent to $(\forall x)(x = a \rightarrow - Fx)$, is also about $\{a\}$. These are useful results but, like Nelson Goodman's discussion, the explications rely essentially on logical syntax. Obviously I have broader concerns here. In addition, Goodman's (1972: 253–5) view is that *Crows are black* is about black things as well as about crows, and that *Cows are animals* is "absolutely about" non-cows. In fact, as Goodman (1972: 258) notes, on his view a statement that is about any class or classes is "absolutely about" each Boolean function of them. The difficulty, apart from these artificial consequences, lies in the basic intuition of Goodman's that besides *Maine has many lakes* being about Maine, *It was Maine that has many lakes* is also about Maine. The latter claim is precisely what I deny (see Atlas and Levinson 1981). Failure to understand the principle distinguishing between these two cases accounts for nearly ninety years of philosophical errors.

Here is a wonderfully candid passage from Keith Donnellan's (1975: 99) essay "Speaking of Nothing":

> Russell's analysis of statements containing definite descriptions and, by extension, ordinary proper names, shows, he believed, that such statements are not really *about*, do not really *mention*, the denotation of the description or the referent of the name. Russell emphasizes this again and again. 'Genuine' names, on the other hand, can somehow perform the feat of really mentioning an individual particular. To try to put much weight on such terms as 'about' would lead us, I think, into a morass. What it is for a statement to be *about* an individual, if that requires any attempt to define *aboutness*, is a question better avoided if we are ever to get on with the problem.

Donnellan's view is quite characteristic of established philosophical opinion on the matter, and it is mistaken about Russell's own view, as I shall show below, unnecessarily extreme in its demand for *definition*, and, for a philosopher as discerning as Donnellan (see Donnellan 1981), unexpectedly oblivious to a rich semantic problem. I have an attitude different from Donnellan's towards this morass, if such it be (I shall argue that it is not). ABOUTNESS is a fundamental topic for linguistics and philosophical logic; so I intend to theorize about it, indeed offer linguistic criteria for it, and not retreat from thinking about (!) the problem because the phenomenon seems too soft according to some pseudo-logical standard. Donnellan, a few paragraphs later, forgets his philosophical scruples and describes what he calls 'the natural, pre-theoretical view' in these words (Donnellan 1975: 99–101):

> if one says, e.g., 'Socrates is snub-nosed', the natural view seems to me to be that the singular expression 'Socrates' is simply a device used by the speaker to pick out what we want to talk ABOUT [*my emphasis*] while the rest of the sentence expresses what property he wishes to attribute to that individual. . . . The 'natural' view . . . seems to generate Russell's budget of puzzles. . . . If I say, 'Socrates is snub-nosed', the proposition I express is represented as containing Socrates. If I say, instead, 'Jacob Horn does not exist' the 'natural' view seems to lead to the unwonted conclusion that even if what I say is true, Jacob Horn, though non-existent, must have some reality. Else what proposition am I expressing? The 'natural' view thus seems to land us with the Meinongian population explosion.

Even the scrupulous Donnellan finds it perfectly easy to formulate the 'natural' view and its alleged consequence through an effortless and apparently intelligible deployment of the very concept of ABOUTNESS that he philosophically disdains. The mistake about Russell that Donnellan, along with nearly all contemporary philosophical opinion, makes is glossing ABOUTNESS as *mention*, i.e. reference, as in the previous quotation but one. It would be historically more accurate in discussing Russell to gloss *mention* as ABOUTNESS. The concept of ABOUTNESS is used by Russell in *The Principles of Mathematics* (1903: 44–5), where he writes:

> [I]n a large class of propositions . . . it is possible, in one or more ways, to distinguish a subject and an assertion [i.e. a predication] ABOUT the subject.

. . . [T]he proposition "humanity belongs to Socrates", which is equivalent to "Socrates is human," is an assertion ABOUT humanity; but it is a distinct proposition. In "Socrates is human," the notion expressed by *human* occurs in a different way from that in which it occurs when it is called *humanity*, the difference being that in the latter case, but not in the former, the proposition is ABOUT this notion. . . . I shall speak of the *terms* of a proposition as those terms, however numerous, which occur in a proposition and may be regarded as subjects ABOUT which the proposition is [*emphasis added*].

It is perfectly clear in this passage that the notion of a term, in the special sense intended by Russell in *The Principles of Mathematics*, is a derivative notion, being explained by the more fundamental notion of propositional ABOUTNESS; *mention* is being explained by ABOUTNESS, not ABOUTNESS by *mention* (i.e. by reference). In the review article 'Meinong's Theory of Complexes and Assumptions", first published in *Mind* (1904), Russell (1973: 53–4) elaborates:

We can and do, apparently, think of objects in a certain relation, without thinking of the relation at all. And apart from inspection, the endless regress seems to prove that this is so; for if not, we should have to think also of the relation of the relation to the terms, and so on, which would make the apprehension of objects in relation impossible. This is, we may conjecture, the reason why common sense and many philosophers regard relations as less real, less substantial in some sense, than their terms.

. . . In some sense which it would be very desirable to define, a relational proposition seems to be *about* its terms, in a way in which it is not *about* the relation.

Students of Russell's logical researches 1900–20 will recognize in Russell's comment on the desirability of a definition of ABOUTNESS a characteristically Russellian attitude: for his present purposes propositional ABOUTNESS is a primitive concept; and as it is for purposes of explanation a basic notion, it would be desirable eventually to reduce it, by an appropriate definition, to even more fundamental notions, though he does not know what such notions are. It is clear, however, that such notions are not 'term', 'mention', or 'reference', which have instead been characterized in terms of 'aboutness'. Russell continues:

And we distinguish between 'A is the father of B' and 'Fatherhood holds between A and B': the latter, but not the former, is ABOUT fatherhood as well as A and B, and asserts, while the former does not, a relation of fatherhood to A and B. Hence, although, at first sight, the difference might seem to be a merely subjective difference of emphasis, it results that there is a real LOGICAL difference [*emphasis added*].

The insight of Russell's last remark seems to have been lost after 1904, absent from Jena, Berlin, Vienna, and even Cambridge for the next fifty years, finally recovered in Oxford about 1954, and since 1964 almost totally ignored.

The evidence for the logical difference between the propositions is precisely Russell's linguistic intuition about the aboutness of statements. Lest it be thought that these remarks of Russell's are merely incidental ones in his philosophical argumentation, it should be noticed that later in his discussion of Meinong Russell (1973: 61) makes essential use of them in criticism of one metaphysical view of truth and falsehood:

[I]f we consider a Relation R between *a* and *b*, we should say, when it is false that this relation holds, that there is no such thing as the relation R between *a* and *b*. But this argument has been already disposed of, by the contention that the being of this relation is not what the proposition '*a* has the relation R to *b*' really affirms.

The being of the relation is not affirmed by the proposition because the proposition is not ABOUT the relation.

From a historical point of view it is interesting to note that in the year following the publication of his article on Meinong Russell espoused in "On Denoting" (1905) precisely the view of falsity that he has here rejected. He (1905/56: 53–4) writes:

If '*aRb*' stands for '*a* has the relation *R* to *b*', then when *aRb* is true, there is such an entity as the relation *R* between *a* and *b*; when *aRb* is false, there is no such entity. Thus out of any proposition we can make a denoting phrase, which denotes an entity if the proposition is true, but does not denote an entity if the proposition is false. E.g., it is true (at least we will suppose so) that the earth revolves round the sun, and false that the sun revolves round the earth; hence 'the revolution of the earth round the sun' denotes an entity, while 'the revolution of the sun round the earth' does not denote an entity.*

* The propositions from which such entities are derived are not identical either with these entities or with the propositions that these entities have being.

Russell's difficulties with his theories of true and false judgements plague him from 1903 to 1920. What is historically fascinating is to see that Russell cannot put together the insights of his paper on

Meinong (1904) with his theory of definite descriptions (1905). In 1904 he abandoned the view that the falsity of *aRb* meant that there is no such thing as the relation *R* between *a* and *b*, on the (correct) grounds that *aRb* is not ABOUT *R*. So *R* is not a "term" in the proposition.

Once he has the theory of definite descriptions, he reverts to an account of relational propositions in which *R* is a "term", which of course produces famous difficulties for a theory of judgement, difficulties lucidly explored by Russell himself. It is as if he were so delighted that 'the revolution of the sun round the earth' does not denote any entity, rather than, as in his earlier view, denotes a non-existent entity that he abandons his earlier insight into the aboutness of relational statements, tacitly expecting that his new theory of denotation will solve whatever problems may arise. The burden is shifted from the semantically more astute conclusion of 1904 concerning the aboutness of propositions to the 1905 theory of the denotation of singular terms. This is a retreat from the Fregean concern with sentences to the Idealist Logicians' concern with terms. Russell's new theory of denotation is grafted on to his less astute assumptions of 1900–2.

Those assumptions were a blend of Peano's logic, G. E. Moore's realist metaphysics, and F. H. Bradley's Idealist theory of judgement. Only a philosopher steeped in F. H. Bradley's *Principles of Logic* (1883) could say, as Russell did in "On Denoting" (1905), that *The earth revolves round the sun* is true if and only if *the revolution of the earth round the sun* denotes an entity, where the relation expressed by the proposition is turned into a definite description, and then by Russell's theory into "an adjective of the real world", to use Bradley's expression from his *Principles of Logic* (1883). Russell's theory of definite descriptions transforms singular judgements into general ones, which is consistent with Bradley's attack on singular judgements. Furthermore, the Fregean account of existence as a second-order property of properties rather than as a first-order property of individuals is consistent with Bradley's doctrine that existence is not a quality. Finally, for a definite description to lack a denotation is, on Russell's theory, for certain properties not to be true of anything. And so, for a relational proposition to be false is for these properties not to be true of anything, i.e. not to be "adjectives of the real world". Russell's theory of denotation saves F. H. Bradley's theory of judgement. In "On Denoting" (1905) Russell abandons the insight into the structure of propositions gained by his reflections on Meinong and reverts to F. H. Bradley's theory of judgement, now thinly disguised by Russell in Peano's logical notation as his new theory of denotation. (For a stimulating but incorrect interpretation that attempts to Russellize Bradley, rather than correctly Bradleyizing Russell, see Manser (1983).)

These passages make it clear that for Russell the justification of the attribution of Being is derived from semantic intuitions of propositional ABOUTNESS. For Russell those intuitions are semantically and epistemologically basic. Those who think that the Russellian problem of negative existence statements arises primarily from the referential vacuity of singular terms ignore the history of Russell's philosophy. Those who believe that the problem of the ABOUTNESS of statements is the same as, or reducible to, the problem of the reference of singular terms confuse presupposing with referring.[19] The problem of existentials arose for Russell out of a fundamental mistake concerning what existential propositions were ABOUT. In subsection *D* below I introduce the linguistic distinctions needed to dissolve the problem in the terms in which Russell originally posed it, something that philosophers have never attempted, much less accomplished. In the next subsection, subsection *C*, I shall briefly look at the problem in the terms in which it arises for Strawson. Once Russell's problem is dissolved, the problem for Strawson can be completely solved. In concluding this historical section, I must mention that in 1903 Russell explicitly commits himself to an analysis of the proposition "*A* is", which, writes (Russell 1903: 49), "holds of every term [*A*] without exception. The *is* here is quite different from the *is* in 'Socrates is human'; it may be regarded as complex, and as really predicating Being of *A*." In later linguistic terminology the topic of the proposition "*A* is" is *A*; the comment is that it has Being. In what follows I shall show that Russell's topic/comment analysis of the existence statement is completely mistaken.

[19] In my view a semantic theory of singular terms will fail unless it makes proper use of the principle that "only in the context of a sentence does a singular term have a reference", because only there does it play the semantic role of referring term, vacuous or not. (I do not exclude one-word sentences, e.g. 'John!'). Interesting discussion may be found in Jonathan Cohen (1980), P. F. Strawson (1980), and Paul Ziff (1977).

C. Strawson's Problem of Negative Existentials

Strawson begins from the basic linguistic intuition that distinguishes asserting from presupposing. Utterances of the form *The F is G*, predicate-negation understandings of *The F is not G*, and *Is the F a G?* each permit a presuppositional inference to *The F exists*. Conventional assertoric and interrogative uses of these sentences carry the presupposition that the singular term *The F* has its denotation. Nothing stops the same consideration from applying to the sentence schema *The F exists*. It would seem that uttering *The F exists*, *The F doesn't exist*, and *Does the F exist?* would each permit a presuppositional inference to *The F exists*. If so, asserting *The F exists* presupposes its own truth, asserting *The F doesn't exist* presupposes its own falsity, and asking the question *Does the F exist?* presupposes the answer *Yes*. Obviously these consequences are absurd.

If, in addition, one adopts a semantic conception of presupposition according to which a statement fails to be true or to be false when one of its presuppositions is false, a statement like *Pegasus doesn't exist* will be predicted to be false if true, false if false, and so necessarily false (if either true or false) or truth-value-less. Since some such statements are contingently true rather than either necessarily false or truth-value-less, one must abandon either the semantic conception of presupposition or the claim that *The F exists* and *The F doesn't exist* presuppose *The F exists*. Those who held to the semantic conception as the only conception of presupposition simply accepted as an *ad hoc* principle that explicit existential statements did not carry existential presuppositions. No explanation of this *ad hoc* principle was ever given. Even if one abandons the semantic conception of presupposition, to avoid this dismal theoretical adhocness, the absurd consequences that I described in the preceding paragraph will remain unexplained.

It is evident that a solution to Strawson's problem with existence statements is to be found by analysing the absurdities mentioned in the preceding paragraph but one: by trying to discover what is wrong with the statement's *The F doesn't exist* presupposing that there is an *F*. In my discussion of (M_S) in subsection *A* I suggested that a statement presupposes the existence of an *F* only if *the F* is a topic-designating Noun Phrase in the statement. The question to ask is: Is *The F doesn't exist*, or *The F exists*, a statement in which *the F* is a topic-designating Noun Phrase, as Russell thought? If it isn't, and Russell was wrong to think that it was, then *The F exists* will not presuppose its own truth, and *The F doesn't exist* will not be truth-value-less. The absurdities in Strawson's account will vanish. In the next subsection I shall answer the question about the topics of explicit existential statements by answering the question about the topic-designating character of singular terms that occur in explicit existential statements. The first question is a metaphysical question; the second question is a logico-linguistic question. The first question does not correctly arise unless the second question has been correctly answered. It is the thesis of this section that heretofore the second question has not been correctly answered by philosophers *except by accident*, notably by Bertrand Russell (1905). Russell got the right answer *without knowing why* it was the right answer. This section explains, among other things, *why* Russell's answer was the right answer, an *explanation* that has not previously been provided. This explanation, if correct, not only fills a long-standing lacuna in logical theory but shows why the failure to answer the logico-linguistic question correctly led the pre-1905 Russell, and many others since, to make wild metaphysical posits.

D. *Statement-Aboutness for Existence Statements*

In a normally stressed statement of *Johnny deceived the girl*, *Johnny* is a topic-designating Noun Phrase (NP), Johnny himself is the topic, and the comment about the topic is that he deceived the girl. In *It was Johnny who deceived the girl*, *Johnny* is not a topic designating NP, so Johnny is not the topic, by which I mean, intuitively, that the statement is not about him. The simple subject-predicate statement that corresponds to the cleft statement is the contrastively stressed statement JOHNNY *deceived the girl*, which is not about Johnny. These data illustrate several generalizations about statements (not sentences, please note):

(i) Topic NPS are unstressed; they receive neither primary nor contrastive stress.

(ii) Leftmost NPS (in surface forms) are not necessarily topic NPS.

(iii) Statements carry presuppositions of the existence of NP designations only if the NPS are topical. (For example, a statement *Johnny deceived the girl* carries a presupposition that the designation of *Johnny* exists, but a statement of *It was Johnny who deceived the girl* does not.)

The last observation was first made by Paul Grice, Geoffrey Warnock, and Peter Strawson (1954, 1964). Strawson's examples were these:

(46) (*a*) *The king of France visited the exhibition.*
(*b*) *The exhibition was visited by the king of France.*

On Strawson's view the former would be neither true nor false, but the latter would be false, if there were no king of France. It would come as no surprise to linguists that passives front a topical NP into the surface-subject position, where with normal stress the NP would be understood as a topic NP. So-called "Topicalization" has a similar effect: *The exhibition, the king of France visited*. Strawson suggested, as had Grice, that the failure of a referring expression to designate an individual would result in a "misfire" in the speech act only when the expression was a topic-designating expression. But these observations failed to elicit systematic discussion from philosophers.

Recently this situation has changed (see Atlas (1981, 1982, 1988*b*), Chapter 2, Section 2, Donnellan (1981), and Hintikka (1986)). Unhappily, Donnellan's discussion is marred by failure to distinguish statement topic from conversation topic in his treatment of Strawson's views. Donnellan's criticism of Strawson's topic analysis depends on comparatives, e.g. *I am taller than the king of France*, which Donnellan takes to lack a truth-value. He assumes that Strawson's conversation-topic analysis would predict falsehood. Even if Donnellan were correct in the matter of truth-value, this criticism of Strawson is too hasty, as it imputes to him a subject-predicate analysis of comparative sentences that is not required by his view and ignores the crucial fact that *topic/comment is a property of statements, not sentences.*[20]

Hintikka (1986: 267) has also given his attention to the Strawsonian puzzle of existence statements:

Strawson and others will have to explain somehow why the existence of a unique present king of France is not presupposed in

(1) *The present king of France does not exist.*

(or, more idiomatically, *There's no such person as the present king of*

[20] An analysis of comparatives is given in Atlas (1984*a*). See also Dresher (1977) and Larson (1988: 14–15 n. 14).

France) whereas it is presumably presupposed by many other assertions in which the definite description occurs, including the famous Russellian example

(6) *The present king of France is bald.*

I take it that the section of this book presently before the reader meets Hintikka's demand. It completes the Strawsonian (1964) account of presupposition. Hintikka (1986: 267–8) continues:

It is nevertheless relevant to note that, in my model, explanation is virtually immediate. The sentence (6) is most naturally thought of as an answer to the question

(7) *Is the present king of France bald?*

whereas (1) is equally readily construed as an answer to (the) following one:

(8) *Is there such a person as the present king of France?*

I think that this won't do at all. First, it is not sentences but utterances that are properly conceived of as question/answer sequences that can be judged deviant or non-deviant, acceptable or unacceptable, or "natural" as Hintikka puts it. Among other things Hintikka is ignoring stress and intonation. (Elsewhere in his paper he is more precise in distinguishing sentences from utterances; remarkably he seems to ignore it just when the distinction matters.) Secondly, Hintikka's *sentence* (6) is *just* as naturally thought of as an answer to the question

(7′) *Who's bald?*

Thirdly,

(1) *The present king of France does not exist.*

is not at all readily construed as a non-deviant reply to the question

(8) *Is there such a person as the present king of France?*

as a moment's attention to one's "ear" will verify, since, fourthly, there are important syntactic, semantic, and pragmatic differences between *The present king of France does not exist* and *There's no such person as the present king of France* (see n. 21). These differences Hintikka has chosen to ignore.

Hintikka (1986: 268) goes on to say:

Now the presupposition of (7) in the precise sense defined in my theory of questions is equivalent to

(9) *There exists such a person as the present king of France*

whereas (8) has in my theory a vacuous presupposition.

This strikes me as an explanation that is so "virtually immediate", to use Hintikka's words, as to be a "vacuous" explanation. One wants a real explanation, now, of the "vacuous" presuppositions of questions like (8). The analysis I give actually explains why a statement of (1) lacks an existential presupposition and so provides a coherent version of Strawson's account. Hintikka's account begs the explanatory question. Having discussed Donnellan's and Hintikka's accounts, I resume the discussion of my own.

In the discussion of the Grice-Warnock-Strawson observations, viz. (iii) above, we have seen that in clefts the syntax marks certain constituents as presuppositional and thus topical. But what John R. Ross (1967) called 'Left Dislocation' also provides a structural indication of the topic status of an NP:

(47) (*a*) *Johnny, he deceived the girl.*
(*b*) *As for Johnny, he deceived the girl.*

if the topic is Johnny, and:

(48) (*a*) *The girl, Johnny deceived her.*
(*b*) *As for the girl, Johnny deceived her.*

if the topic is the girl. Statements of these last sentences would be felicitous in a context only if it would be appropriate in the context to predicate something of the girl.

The linguist Jeannette Gundel (1977) makes a number of subtle observations about statements with topical NPs. Gundel takes the following Question-Answer sequences to be acceptable:

(49) (*a*) *What about Johnny? Johnny is such that he deceived the girl.*
(*b*) *What about Johnny? As for Johnny, he deceived the girl.*
(*c*) *What about Johnny? Johnny, he deceived the girl.*
(*d*) *What about the girl?* JOHNNY *deceived the girl.*
(*e*) *What about the girl? It was Johnny who deceived the girl.*
(*f*) *What about the girl? As for the girl, Johnny deceived her.*
(*g*) *What about the girl? The girl is such that Johnny deceived her.*

She takes the following Question-Answer sequences to be unacceptable:

(50) (*a*) *?What about the girl? Johnny is such that he deceived the girl.*
(*b*) *?What about the girl? As for Johnny, he deceived the girl.*
(*c*) *?What about the girl? Johnny, he deceived the girl.*
(*d*) *?What about Johnny?* JOHNNY *deceived the girl.*
(*e*) *?What about Johnny? It was Johnny who deceived the girl.*
(*f*) *?What about Johnny? As for the girl, Johnny deceived her.*
(*g*) *?What about Johnny? The girl is such that Johnny deceived her.*

Analogously, we have:

(51) (*a*) *As for Johnny, he deceived the girl.*
(*a'*) *As for the girl, Johnny deceived her.*
(*b*) *?As for Johnny,* HE *deceived the girl.*
(*b'*) *As for the girl,* JOHNNY *deceived her.*
(*c*) *?As for Johnny, it was he who deceived the girl.*
(*c'*) *As for the girl, it was Johnny who deceived her.*

The last pair of statements indicates that the clefted NP is not a topic-designating NP in cleft statements. So, on Strawson's account, even if stating *The king of France visited the exhibition* misfires, and no true or false attributive singular assertion is made, a statement of *It was the king of France who visited the exhibition* would be false, which intuitively it is (see Atlas and Levinson (1981), Atlas (1981), Levinson (1983: 217–22), J. McCawley (1981: 241), and Chapter 2, Section 2).

The sentence 'John runs' may be uttered with normal stress and intonation in reply to a question 'What about John?' Normal stress in such a statement means that the primary stress falls on the last word of the sentence, viz. 'runs'. The 'about'-question specifies the topic John, about which the statement that John runs makes its comment. Equally acceptable replies to the question would be the left-dislocated sentence 'John, he runs' and the 'as-for'-sentence 'As for John, he runs'. In general, we have the following:

(C) *Criteria of Noun Phrase Topicality*
(1) If an expression NP is a topic Noun Phrase in a statement, the statement is a linguistically acceptable answer to the question *What about* NP?

(2) If an expression NP is a topic Noun Phrase in a statement, the *as-for-NP* transform of the statement is linguistically acceptable (in the same contexts) and makes the same statement.

(3) If an expression NP is a topic Noun Phrase in a statement, the expression never receives primary stress.

(4) (*The Strawson/Grice Condition*)
The existence of a referent of NP is presupposed in making a statement only if NP is a topical subject, i.e. a topic-designating expression, in the statement.

In light of these criteria, we may now analyse explicit existence statements.

First let us consider criteria (C1) and (C3). With normal intonation and stress, to what question is the statement of *John exists* a felicitous answer? I think it is *Who/What exists?* (and more generally, *What about what exists?*). The topic is: what exists, according to (C1). This conclusion is reinforced by the observation that the primary stress in an informative statement of *John exists* is carried by *John*, which is what one would expect when the statement is used in a context in which it has not already been established that John exists.

There are spontaneous linguistic data that bear on this point. During dinner in Cambridge with Stephen Levinson, my friend Mikael Dolfe, a Swedish journalist and novelist, who had not met Levinson before, asked him whether he had a family, since they were not in evidence in the house. Levinson, whose wife Penny Brown and son Nicholas were in Australia at the time, replied, "I have a family—they're not here, but they EXIST."

Levinson's emphatic stress on 'exist' made sense in the context because he had already established the existence of his family by asserting *I have a family*. So his *they*, the topic NP in his utterance, did not receive primary stress, which is predicted by (C3). In his assertion of their existence, viz. *I have a family*, normal stress gives *a family* primary stress. In English we do not typically use 'My family exists', because normal stress patterns in English put primary stress at the end of the sentence, while the comment rather than topic role of *family* in *My family exists*, which is the same comment role as in *I have a family*, would put primary stress at the beginning of the sentence. We avoid this difficulty in stress placement by saying *I have a family*, where the two regularities agree in placing primary stress.[21] By criteria (C1) and (C3) the topic of a non-emphatic statement *John exists* THAT CARRIES NEW INFORMATION IN THE CONTEXT is not John but rather what exists. Now let us consider criterion (C2).

Careful attention to the following two pairs of statements, when stress and intonation for each member of the pair make each appropriate in the same contexts, will show that the first pair are paraphrases, while the second pair are not:

(52) (*a*) *John exists.*
(*b*) *As for what exists, John does.*[22]

(53) (*a*) *John exists.*
(*b*) *As for John, he exists.*

The correct pairing for (53*b*) is the contrastively stressed:

(*a*′) *John* EXISTS. (e.g. *John doesn't* LIVE, *John* EXISTS.)
(*b*) *As for John, he exists.*

The intuitively correct pairing (52) suggests by criterion (C2) that *John* is not a topic NP in a non-emphatic, informative statement of *John exists*; the topic is: what exists. Of course, if *John* is not a topic NP in a statement of *John exists* the existence of John is not presupposed in making the statement, according to criterion (C4). And it surely is the case that in a non-emphatic, informative assertion of *John exists* John's existence is asserted, not presupposed.[23]

[21] For a recent discussion of the linguistic connection among statements of existence, location, and personal possession in English and, it seems, in many human languages, see R. J. Lizotte, "Universals Concerning Existence, Possession, and Location Sentences", Ph.D. Diss., Brown Univ., 1983.

[22] For those with delicate ears, some variants of this are: (i) *As for existing, John does.* (ii) *As for the existing, John is one.*

[23] Over the years I have asked for judgements on these pairs from over fifty native speakers of American English who are not professional philosophers or linguists. No speaker has balked at making the paraphrase judgement because he or she thought the sentences were "unnatural". Every experimental subject has expressed the preference for the paraphrase that my theory predicts he should express, without exception. I take this to be somewhat encouraging empirical evidence for the position that I outline here on the topic/comment structure of existence statements. Native speakers also report the kinds of context in which the pair-matches diverge from their preferred match of the paraphrases. These are rhetorically distinctive uses of *John* EXISTS to reaffirm, emphasize, or reject the denial of a prior commitment to John's existence. But even a philosopher of language, linguistically corrupted as he is, might have a theoretical reason to agree with me and with the empirical evidence of native speakers' judgements. If he has read Frege or Strawson, he might imagine that a statement of *As for John, he exists* involves a presupposition of John's existence in the anaphoric use of 'he' that is absent from *John exists*, i.e. a referential redundancy in the former that is absent in

Parallel arguments support the same conclusion for explicit negative existence statements. A quick argument to this effect is easily formulated: presuppositions are invariant under main-verb negation. If *John does not exist* were to presuppose John's existence, so would *John exists*. But the latter does not; so neither does the former. It is also easy to show directly from applications of our criteria that *John* in a non-emphatic, informative statement of *John does not exist* is not a topic NP. By criterion (C4), if it is not a topic NP the existence of its designation is not presupposed in a statement of the sentence.

In addition to these criteria there is a paraphrase argument that I find quite suggestive. Consider that in the non-existence clefts

(54) *?It's nothing that is Pegasus.*

and

(55) *It's not Pegasus that exists.*

in which *Pegasus* is respectively topical and not topical, only the latter, non-topic, non-existence statement is grammatically acceptable. The alternative negative cleft

(56) *It's Pegasus that doesn't exist.*

on some theories entails and on all theories at least implicates that only Pegasus does not exist (Atlas and Levinson (1981), Horn (1981)). This is an incorrect prediction except in contexts that delimit a list of names or descriptions, all of which, excepting *Pegasus*, are singular terms that designate existents. The closest of the three negative clefts (54), (55), and (56) to

(57) *Pegasus does not exist.*

is, therefore, (55), the topic of which is: what exists.

I have one further argument. My analysis of the non-topic character of the NP in informative, non-emphatic existence statements is confirmed by the unaccepability, and peculiar redundancy, of

the latter. This reflection might give the philosopher a reason to prefer *As for what exists, John does* and its ilk as better paraphrases of non-emphatic, informative statements of the sentence 'John exists'. Every experimental subject, though not every philosopher of language, as I have been disappointed to discover, realizes that the paraphrase judgement is a matching of statements, not sentences, that stress and intonation are directly relevant to the judgement, and that they are being asked to judge whether they would use the sentences, with the appropriate stress and intonation contour, in the very same contexts.

(58) *?As for what does not exist, Pegasus does (not exist).*

which makes it unavailable as an acceptable topic/comment paraphrase of (57), viz. *Pegasus does not exist*. The latter can best be paraphrased by

(59) *As for what exists, Pegasus doesn't.*[24]

The topic is: what exists. Thus *Pegasus* is not a topic NP in non-emphatic, informative statements of the sentence 'Pegasus does not exist'.[25]

E. *Consequences of My Analysis*

(1) Quite aside from its mathematical advantages, Russell's theory of definite descriptions has seemed intuitive to many mathematicians, logicians, and philosophers. Even if they accept Peter Strawson's criticism of Russell, most logicians and philosophers standardly regard the choice between Russell's classical bivalent theory and a truth-value-gap theory as a pragmatic, not a principled, choice (see Chapter 3, Section 1). But *why*, despite Strawson's criticism, does Russell's theory have some intuitive linguistic appeal?

The standard answer is: because Russell's theory gets the truth-conditions right (or almost right, or right enough for deductive purposes) and because, as Russell (1905/56: 47–8) himself claimed, the theory preserves the Indiscernibility of Identicals, the Law of Excluded Middle, and, most importantly, solves the semantic puzzle that a nonentity can be the subject of a proposition, and in

[24] Again, variants are: (i) *As for existing, Pegasus doesn't*. (ii) *As for the existing, Pegasus isn't one*.

[25] I have eschewed, intentionally, a discussion of the topic/comment analysis of complex sentences with intensional verbs, e.g. 'believes', etc. I have treated the matter in Atlas (1982). In that essay I discuss, among other things, Quine's (1976: 185–96) famous argument in "Quantifiers and Propositional Attitudes" against Quantifying In. His argument rests upon a false assumption about the aboutness of *belief*-statements. In particular, the concluding inference in Quine's (1976: 187–8) argument, from the failure of Ortcutt-aboutness in belief-contexts to the failure of 'Ortcutt'-referentiality in 'believes'-contexts, relies on an assumption, clear in his discussion, that if a singular term occurs referentially in a statement, the statement in which it occurs is ABOUT the referent of the singular term. This assumption is a natural one if the language is an unnatural one, e.g. the language of first-order quantification theory. However, in making this assumption about English Quine has fallen victim to confusing presupposing with referring. The assumption in question, discussed in subsection *A*, is false.

particular the semantic puzzle that it must always be contradictory to deny the being of anything. The solution, as Russell (1919*b*: 170) later phrased it, was that *The F does not exist* is true when *the F* describes nothing real, rather than when it describes something unreal, a case of not describing (not denoting) any entity rather than a case of describing (denoting) a nonentity. Nonentities are not the subjects of propositions, and it is not contradictory to deny the being of anything. Nevertheless, none of this, and especially not the standard remark about the truth-conditions, answers my question: *Why* does Russell's theory have some intuitive, linguistic appeal?

Russell (1956: 250) himself, in his 1919*a* "The Philosophy of Logical Atomism", knows the answer, but he does not know that it is the answer. He writes, ". . . when I say 'The author of *Waverley* exists' I am not saying anything about the author of *Waverley*".

My analysis shows why there is some intuitive, linguistic plausibility in Russell's theory. He gets ABOUTNESS right! So the logical structure of an existential statement is not radically different from a grammatical (viz. its topic/comment) structure. It is false to say, as Russell (1919*b*: 169) did, that it is for "want of the apparatus of propositional functions" that "many logicians have been driven to the conclusion that there are unreal objects". Rather, it is true to say, as Russell immediately did, that it is "argued, e.g. by Meinong, that we can speak ABOUT the golden mountain, the round square, and so on; we can make true propositions of which these are the subjects; hence they must have some kind of logical being". This is the argument that led Meinong and Russell to the conclusion that there are unreal objects, even when what was being said was that the golden mountain does not exist. I have argued that in a non-emphatic, informative statement of *The golden mountain does not exist* the speaker, in a perfectly coherent sense of 'about', is NOT speaking ABOUT the golden mountain. The golden mountain is NOT the "subject" of the existential proposition.

Of course, Russell's logical form is different from surface form, but his logical form is consistent with my analysis of the topic/comment structure of explicit existential statements. On Russell's view *Spies exist* is about a propositional function "*x* is a spy" and says of it that it is instantiated. But we need not interpret the quantification in Russell's metalinguistic manner. $\exists x(Spy\ x)$

might be viewed as about the domain of quantification *D*, what exists, asserting of it that it contains at least one spy. On this rather natural view of the matter logical analysis and linguistic analysis agree. Logical grammar does not revise linguistic grammar. It merely reports it.

(2) Another virtue of my analysis is that it *explains* the anomalies discussed in subsection *C* completely. As Strawson (1964) observed, a statement *A(t)* presupposes *t exists* only if *t* is a topic *NP*. As I have shown in subsection *D*, in a non-emphatic, informative statement of *t exists*, *t* is not a topic *NP*. Thus the existential statements *t exists* and *t does not exist* do not presuppose *t exists*. The anomalies in Strawson's basic conception of presupposition are removed by a correct linguistic analysis of the topics of explicit existence statements.

(3) Russell's and McZ's problems of non-being are linguistic pseudo-problems. In Russell's case he thought, prior to 1905, that statements of *Pegasus exists* were about Pegasus, i.e. he thought that in those statements *Pegasus* was a topic Noun Phrase. Similarly, in statements of *Pegasus does not exist*, he thought that *Pegasus* was a topic Noun Phrase. It is not, and so those statements are not about Pegasus. They are about what exists, or, as Quine would have to say: they are about everything. THE METAPHYSICAL PROBLEM OF PEGASUS'S BEING SIMPLY DOES NOT ARISE FROM THE MEANING OF NEGATIVE EXISTENCE STATEMENTS.

(4) As McZ's puzzle of non-being shows, Quine's account of Russell's problem with negative existentials fails to deal with the underlying semantic problem. The problem was not Russell's confusing meaning with naming. (Russell never confused linguistic meaning with naming, as anyone who has read Section 51 of *The Principles of Mathematics* (1903) should know.) The problem was Russell's mistaking what negative existence statements were about, or to put it in his other terminology, mistaking what the subjects of negative existence propositions were. From my point of view, Russell mistook what were the topic-designating Noun Phrases. Having made that linguistic mistake, the metaphysical mistake as to what negative existence statements were about was logically inevitable.

The point that has been missed by Quine, Goodman, and most other philosophers is this: *Not every singular term in a statement will designate what the statement is about.* So the statement need not presuppose the existence of the referent of every singular term that it contains.[26]

The philosophical problem of negative existence statements has been derived entirely from a linguistic mistake.[27]

I have called this chapter a case study in philosophical linguistics because in it I wished to show how novel applications of linguistic theory to classic problems in philosophical logic and philosophy of language can revise, indeed radically transform, our view of those problems. I have wished to examine the standard dogmas that we have inherited from Russell and Strawson and resolve finally the stalemate between their views. Russell (1905) and Strawson (1950, 1952) were both wrong, though each was in part right. A good theory that overcomes a stalemate should preserve the good parts of the antecedent theories and jettison the bad parts; in so doing it will, almost always, need to reconceptualize the problems, introduce new vocabulary, solve long-standing anomalies in the old theory, and raise new problems for further investigation. For me at any rate the Atlas-Kempson thesis of the sense-generality of 'not' sentences has led to new questions about presupposition sentences and truth-value gaps; that in turn has led to a reconsideration of Strawson's notion of a statement and thus to an examination of the linguistic notion of statement topic/comment. As a consequence I have removed a long-standing anomaly in Strawson's conception of presupposition, viz. the anomaly of the missing presuppositions of existence statements.

I realized, after first formulating the views of this section, that

[26] Throughout this section I have limited my discussion to extensional statements. The point just made is trivial for opaque contexts as standardly understood.

[27] Ray Jackendoff (1983: 88–91) has argued, and I agree, that the English word 'is' is unspecified for the feature opposition [OBJECT]/[CONCEPT], which Jackendoff calls 'TOKEN/TYPE opposition'. 'Is' in 'Clark Kent is a reporter' and in 'Clark Kent is Superman' is not ambiguous between an 'is' of predication and an 'is' of identity. Jackendoff concludes that the "verb 'be' surrounded by two Noun Phrases has only a single reading, which we may call 'BE(*x*, *y*)', capable of comparing either two [TOKENS] or a [TOKEN] and a [TYPE]", i.e., ignoring a nicety in Jackendoff's discussion, the linking of two singular terms or the linking of a singular term with a General term (Jackendoff 1983: 90). Hence, 'is' is sense-unspecified in meaning between the specifications "is identical to" and "is an instance of", not ambiguous between them. Thus another traditional ambiguity doctrine, that 'is' is ambiguous in English among an 'is' of identity, an 'is' of predication, and an 'is' of existence, is simply false philosophical dogma.

Strawson (1954, 1964) had himself provided part of the materials for the solution. I am sure that unconsciously I made use of those ideas, which I had studied some years before. At first the embarrassing question was why it took me so long to follow out Strawson's hints, but it turned out that I had to solve a semantic problem first, the problem of negation, before I understood the significance of Grice's and Strawson's observations on topic/comment. And I would never have discovered the importance of topic/comment in the history of Russell's philosophy if Michael Dummett had not recommended my studying Russell's (1903) *Principles of Mathematics*. I was then able to make sense of the remarkable discontinuity in Russell's account of the truth and falsity of relation statements from 1904 to 1905, when by abandoning Meinong's ontology he also abandoned his own Meinongian linguistic insight into the topic/comment character of relation statements. And then I was able to see that Quine's "On What There Is" was a false diagnosis of Russell's problem of existence statements.

Many of my claims in this chapter will be controversial, but philosophers and linguists will probably worry most about the thesis that 'not' sentences are sense-unspecified for scope. As I have claimed that a good theory should remove long-standing anomalies, I now want to show that the thesis removes a very puzzling, very ancient anomaly in logical theory.

4. The Anomaly of the Ambiguity of Negation in Logical Theory

One traditional problem of negation was nicely formulated by Sir Alfred Ayer (1954: 46–7) when he wrote:

> If I say that the Mediterranean sea is blue, I am referring to an individual object and ascribing a quality to it; my statement, if it is true, states a positive fact. But if I say that the Atlantic is not blue, though I am again referring to an individual, I am not ascribing any quality to it; and while, if my statement is true, there must be some positive fact which makes it so, it cannot, so the argument runs, be the fact that the Atlantic is not blue, since this is not positive, and so, strictly speaking, not a fact at all. Thus it would seem either that the apparently negative statement is somehow doing duty for one that is affirmative, or that it is made true, if it is true, by some fact which it does not state. And it is thought that both alternatives are paradoxical.

If one asks by virtue of what is the true statement that the Atlantic is not blue made true, one famous way between the horns of the dilemma Ayer has posed is to reply "a fact", i.e. a "negative fact", a language-independent and thought-independent entity distinct from the fact that the Atlantic is grey, or the fact that it is green, etc. The statement that the Atlantic is not blue is neither a surrogate for a statement that the Atlantic is grey nor is it made true by some fact that it does not state. It is made true by a negative fact that it does state.

The counter-intuitive manner of this brave solution is characteristic of Bertrand Russell, who had the courage of G. E. Moore's realist convictions in the second decade of this century. What clinched the point for Russell was that if it is false that the Atlantic is blue, it is false because of a fact in the world, "the" very fact that makes it true that the Atlantic is not blue. The affirmative proposition is refuted by a fact. So the negative proposition is verified by, and thus on Russell's view does correspond uniquely to, that fact. Otherwise, Russell thought, one would neither be able to explain why it is false that the Atlantic is blue, nor true that the Atlantic is not blue (Russell 1919*a*/56: 211).

Despite all of the discussion of Russell's logical atomism, one aspect of Russell's view seems never to have been remarked upon. If the negative fact that the Atlantic is not blue is distinct from the fact that the Atlantic is grey, that it is green, and so on, for various predicates incompatible with 'blue', it suddenly takes on a feature that Locke would have recognized: it is an "abstract" fact, the metaphysical, sentential analogue of the "abstract" predicate 'non-blue' and "abstract Idea" non-blueness.[28] Without his realizing it Russell's account of negative facts implied that negative facts were "abstract" and "imperfect" entities in roughly Locke's senses.[29] A Lockean way to put this would be that the Idea non-blueness is "framed" from the recognitions that greyness, greenness, etc. each

[28] To forestall confusion I must mention that I use the expression 'non-blue' as the logical complement of 'blue', not as a contrary to 'blue' and not as a predicate whose extension is restricted to coloured things. One must choose some terminology for this notion, and the alternative 'not-blue' has never appealed to me. In any case I think I am being faithful to Russell's intention in my interpretation of him. Historians of logic and of linguistics have differed in their terminology, and readers of other works should be aware that the literature is afflicted by terminological diversity. See Horn (1989).

[29] See my discussion of Locke in Ch. 1.

differs from blueness and "cannot be comprehended under that name". Thus the negative sentence is a Russellian expression of an "abstract" negative fact; it may also be thought of as a Lockean attribution of an "abstract" negative property to an individual.

Locke's account of "abstract" General terms and Russell's conception of an "abstract" negative fact together suggest that negation, which is a logical constant in the formal language of elementary logic, is in natural language semantically more like a predicate, like a General term, than a logical constant.[30] It would then be possible for 'not' to be sense unspecified for, rather than ambiguous between, exclusion- and choice-negation interpretations of negative sentences.

The existence of an "abstract" Lockean negation, an "imperfect" negation made "perfect" in different ways to give exclusion- and choice-negation interpretations of utterances of a sentence whose semantic character is sense-non-specific beween those interpretations is, at the least, a coherent if novel semantic possibility. It would also solve a long-standing puzzle about the semantics of negative sentences in two millennia of logical tradition.

The puzzle is this. If we negate an atomic sentence in elementary logic, notably an atomic sentence consisting of a one-place predicate symbol, the usual parentheses, and an individual constant, the distinction, both syntactic and semantic, between sentence and predicate negation collapses. Correspondingly, the distinction between the truth of a statement of '$-Fa$' and denying 'F' of a, or between a negative fact (it's not being that a is F), and

[30] This is an idea that seemed plausible to me in 1970 on first studying Klima's classic analysis of negation (Klima 1964). See also Akmajian and Heny (1975). Michael Luntley (1988) reports that in the summer of 1987 Michael Dummett made the claim that negation is not a logical constant, which should be distinguished from my claim that English 'not' (and any free negation morpheme in other natural languages) is not a logical constant. Following Gentzen's suggestion that introduction rules for the constants fix their meanings, the introduction rules are taken by Dummett to be "self-justifying". Further principles of logic are justified if some proof of them employs only self-justifying principles. Luntley (1988: 107, 9–10) reports Dummett to have claimed that negation is not a logical constant because its sense cannot be explained by self-justifying principles. Though Luntley himself believes that *classical* negation is not a logical constant, he believes that a negation operator that is proof-theoretically intuitionistic is *the* negation constant. I leave for another occasion a discussion of the connections between my problem in the semantics of natural language and Dummett's (1973) problem in the epistemology of logic. Philosophers and linguists will find Horn (1989) to be an indispensable historical and critical study of negation, with an encyclopaedic treatment of linguistic data.

the attribution of a negative predicate to *a* (*a*'s being *non-F*), collapses.

The traditional view of this matter in logical theory is simply to accept it as a fact of nature; it is merely a semantic peculiarity of the negations of atomic sentences. Negating atomic sentences results in the equivalence of a sentence operator 'not' with a predicate operator 'not'. Indeed, this equivalence in truth-conditions will distinguish negating atomic sentences from negating molecular sentences and from negating sentences with complex singular terms. The scope distinctions present in the case of complex, molecular sentences simply collapse when confronted with the simple syntax of an atomic sentence. The phenomenon distinguishes atomic sentences from molecular ones. The negations of atomic sentences experience semantic collapse. My puzzle is: Why are the negations of atomic sentences so peculiar in this respect?

Let's think about this in broader terms, not merely in terms of the syntax of a formal language adequate for elementary logic. Let's think about this for ordinary language. Suppose one started from the assumption that, in complex sentences in English, negation yielded structurally or lexically ambiguous sentences. Consider an elementary, simplex sentence with designating singular terms, and then negate it: '*a* is not *F*'. This 'not', which in complex sentences apparently has the power to create interpretations that we explain by a structural ambiguity or by the existence of more than one lexeme 'not_1', 'not_2', etc., either now fails to exercise this syntactic or semantic power or, if it does, results in truth-conditionally equivalent interpretations. If 'not' fails to exercise its power, one wonders why. If the ambiguity remains in 'John Walter is not ugly', isn't it interesting that two distinct senses of an ambiguous sentence should turn out to have equivalent truth-conditions? This puts the simple negative sentence on a par with J. L. Morgan's 'Someone is renting a house', normally thought to be an unusual case, obviously distinct from 'They saw her duck' or from 'He stopped at the bank'. The puzzle is: Why does the ambiguity introduced by 'not' collapse into equivalent propositions? Why should semantic collapse distinguish simplex from complex sentences?

Logicians and linguists have assumed that negation in natural language is ambiguous, either syntactically in complex sentences or lexically in simplex ones. According to Russell (1919*b*: 179; and Russell and Whitehead 1910/27: 69), if a proposition is derived from the substitution of a definite description in a negative-propositional function symbol, e.g. '*x* is not bald', the description is said to have a primary occurrence in the proposition. If the substitution of the description in a propositional function symbol gives only part of the proposition (e.g. the substitution of 'the present king of France' for the variable in '*x* is bald' gives 'the present king of France is bald', which is only part of the proposition 'the present king of France is not bald') the description is said to have a secondary occurrence in the proposition. Russell himself viewed the proposition 'the present king of France is not bald' as ambiguous (Russell 1919*b*: 179; and Russell and Whitehead 1910/27: 69). Corresponding to the secondary occurrence of the description is the derivation of the proposition from a simpler affirmative proposition; corresponding to the primary occurrence of the description is the derivation of the proposition from a negative propositional function. Russell's technical vocabulary indicates that some occurrences of descriptions are more "basic" than others, and the truth-value of the negative proposition depends on the kind of occurrence if the description fails to have a denotation. But, when the definite description has a denotation Russell's two "readings" are ACCIDENTALLY equivalent, as he observes (Russell and Whitehead 1910/27: 69). Even though modern philosophers since Locke have come to recognize—though not under this description—the sense-generality of General terms, recovering something of the logical tradition of late Scholasticism with which Locke was intimately familiar, a logical term like 'not' has been assumed to be syntactically ambiguous in complex sentences between external negation and internal negations, or lexically ambiguous in simplex sentences between exclusion and choice negations. It is to be distinguished, on this view, from a General term like 'neighbour', which is not ambiguous between readings "nearby dweller who is male" and "nearby dweller who is female" but instead has a sense that is grammatically unspecified for gender.

My hypothesis concerning 'not' is unorthodox: the contribution to the sense of a sentence in which 'not' occurs results in a univocal sentence. The standard ambiguity account simply cannot explain why simplex sentences with designating singular terms are semantically special cases in which an ambiguous 'not' curiously collapses into equivalent propositions. Rather than view this equivalence of a sentence-modifier 'not' and a predicate-modifier 'not' as a peculiarity of the negations of atomic or simplex sentences, one that distinguishes them from the negations of molecular or complex sentences, I suggest that in ordinary, natural language no such peculiar distinction exists to mark the difference between negations of simplex and complex sentences. In no simplex English sentence is negation ambiguous between exclusion-negation and choice-negation interpretations. 'Not' is not ambiguous between, but sense-unspecified for, interpretations with differing scopes. The sentences are scope-non-specific. My hypothesis resolves the explanatory anomaly created by the traditional ambiguity account of negative sentences by denying the ambiguity that creates it. There is no ambiguity, and the anomaly is explained away. 'Not' is sense-invariant; the variant understandings of the utterances of negative sentences as exclusion negations or choice negations, predicate negations or sentence negations, must be *constructed* from the contexts in which the utterances occur.

The gifted Jacob Bronowski (1978: 21) once posed what he understood to be "*the* problem of the end of the twentieth century for anyone who seeks to examine all aspects of nature: How are we able to exercise fine discrimination when the units with which we work (like the rods and cones in the eye . . .) are so coarse?" Or to put it in a Kantian style, How are our "fine" interpretations of our "coarse" representations possible?

I have suggested that the coarseness of the units extends from the cells of our nervous system to the pictorial and semantic representations of our art and language, which, like the rods and cones, in combination, make right depictions and right descriptions of a world possible. Some of our "fine" interpretations are not semantically possible; they are only pragmatically possible. Pragmatic inference without sense-generality is blind, but sense-generality without pragmatic inference is empty. This essay seeks to contribute to a solution to "the problem of the end of the twentieth century for anyone who seeks to examine all aspects of nature", at least in so far as that problem arises in the linguistic aspect of nature. To a sketch of that solution I now turn.

4

Understanding Utterances: Figuring Out what Sentences "Portray"

1. Truth and Sentence-Meaning[1]

Philosophers of language often work on technical problems that, they hope, have larger philosophical consequences—e.g. research on the semantics of adverbial modifiers or of propositional-attitude verbs supports a particular philosophical programme in the theory of meaning. This is the programme of those who, like Frege, the early Wittgenstein, Donald Davidson (1967*b*), or David K. Lewis (1972), identify the core of a theory of meaning with a Tarski-like theory of truth and the meanings of statements with their truth-conditions or with the propositions (sets of possible worlds) that they express. Recent linguistic and philosophical work by Ruth Kempson (1975) and myself (1975*a*/*b*, etc.) that I discussed in Chapter 3 casts doubt on a realist, truth-theoretic theory of meaning. The linguistic research has suggested that negation in sentences like 'The king of France is not wise', 'John did not butter the toast in the kitchen at midnight', and 'Mary did not realize that she had failed the examination' is not "scope ambiguous" but rather sense-non-specific for scope. The arguments employ tests discussed by John R. Ross, Noam Chomsky, George Lakoff, James McCawley, Arnold Zwicky, and Jerrold Sadock to distinguish sense-generality from ambiguity. These tests include the assertion and denial of privative opposites, semantic differentia, difference of meaning, transformational potential, and the conjunction test (Zwicky and Sadock 1975). In this section I shall support my sense-generality account of negation by showing that there are intuitive linguistic judgements for which only sense-generality, not ambiguity, can be an explanation. I shall then argue that the sense-generality of negative presuppositional sentences makes it impossible to identify the semantic representation

[1] This section contains, in revised form, portions from Atlas (1978), © 1978, by The Mind Association.

of the sentence with a logical form as conceived by Gilbert Harman (1970) and Donald Davidson (1970), thus refuting the Davidson–Harman hypothesis that underlying structure is logical form. (For modifications in his view see Harman (1973).) It is also impossible to account for sense-generality by Richard Montague's (1970), Robert Stalnaker's (1972), or David Kaplan's (1979) concept of the *meaning* or *character* of the sentence. A theory that is consistent with the linguistic claim that the negative sentences are sense-unspecified for scope and that can explain linguistic intuitions about them must borrow principles from an anti-realist theory of meaning, because the focus of anti-realism in the theory of *meaning* is a theory of *understanding*.

In the theory of meaning the difference between the realist and the anti-realist may in part be described as follows. The realist believes that

(R_1) an explanation of the meaning of a statement *S* consists in giving truth-conditions: the conditions for *S* that are necessary and sufficient for its truth (and falsity);

(R_2) an explanation of the meaning of a predicate *P* consists in giving application-conditions: the conditions for *P* that are necessary and sufficient for *P* to apply correctly to an individual *x* (and incorrectly to *x*).

He also holds the following correspondence theses:

(C_1) For any statement *S*, there is in the world something in virtue of which it is true or false;

(C_2) for any predicate *P* and individual *x*, there is in the world some property of *x* in virtue of which *P* correctly applies to *x* or fails to apply to *x*.

On the realist's view the *understanding* of a statement consists in knowing what has to be the case for the statement to be true. Since the truth-condition need not be a condition that a fallible, limited human being can recognize as obtaining when it does obtain, or for which he has an effective procedure for determining whether it obtains, the meanings of his statements are such that their truth-values are, in general, independent of whether he knows or can know what their truth-values are.[2]

[2] This formulation, given in Atlas (1978), is intended to indicate a separation of two semantic issues: whether a sentence has a *determinate sense* and whether a

statement has a *determinate truth-value*. In Tennant (1981) and Luntley (1988) there is a parallel metaphysical distinction between whether a thought has an objective content ("the objectivity of content") and whether a content has a determinate truth-value ("the objectivity of truth"). The former claims that our thoughts about individuals in an objective spatio-temporal order make sense. This implies that some contents *can* be true, in a recognition-transcendent sense of 'true', whether or not we can verify them. The latter claims that a content has a determinate truth-value whether or not we have knowledge of that value. My concerns here are semantic, not metaphysical, and are directed to Crispin Wright's (1987) question about "the objectivity of meaning", though from a very different perspective. I have been claiming that the sense of 'not' is not "determinate" because 'not' is general in sense. This section elaborates and justifies this claim further.

Wright (1987: 7) states that a "believer in the objectivity of meaning of a certain class of statements will hold that their meanings, in conjunction with appropriate states of the world, can determine their truth-values independently and in advance of any opinions we may form". One who accepts an evidence-transcendent notion of truth and the equation of the meaning of a sentence with its truth-conditions will, of course, believe in the objectivity of a sentence's meaning. Since in this book I give novel grounds for rejecting this equation (I am a *semantic* anti-realist), I am prepared to reject this "objectivity of meaning" if no semantically anti-realist form of objectivity is possible.

In another characterization of the objectivity of meaning, Wright (1987: 5) states that "the meaning of a statement is a real constraint, to which we are bound, as it were, by contract, and to which verdicts about its truth-value may objectively conform, or fail to conform, quite independently of our considered opinion on the matter". Wittgenstein's (1958, 1978) philosophy of mathematics certainly puts in question the objectivity of meaning of mathematical sentences as a "real constraint to which we are bound by contract", what Wright (1987: 6) calls 'the normativity of meaning'. Wright (1980, 1987: 6) suggests that some grounds that put in question the evidence-transcendent notion of truth will also put in question the objectivity of meaning. Even though I reject the truth-conditions notion of sentence-meaning, it seems that I do not *thereby* abandon an evidence-transcendent notion of truth. I merely abandon *the equation* of truth-conditions with sentence-meaning. So, in Wright's view, my semantic anti-realism does not put in question this last kind of objectivity of meaning, even though it *does* put in question his first kind, the kind in which meanings, and the world, *determine* truth-values independently of *any* of our opinions. This is Wright's (1987: 28) notion of objectivity of meaning in which meaning yields up determinate truth-conditions. Sense-general sentences are not meaningful in that kind of way. Indeterminate (and I do not mean 'vague'—see chapter 2, n. 19) senses need not yield determinate truth-conditions, and so need not determine truth-values independently of any of our opinions (see further chapter 4, sections 1 and 3). The upshot is that Wright's (1987) different formulations of the objectivity of meaning fail to provide a well-defined characterization of "the objectivity of meaning" for sentences of a natural language. Needless to say, I am *not* addressing here the Kripke-Wittgenstein sceptical problem: how behaviour or thought "fixes" the meaning of a sentence at all (Kripke 1982; Wright 1987: 149). I am, for the sake of my argument here, being a "naive realist" about the meaningfulness of a sentence. My problem is the *kind* of meaning a meaningful, sense-general sentence has. The kind of meaning that it has makes nonsense of Wright's characterization of the objectivity of meaning.

In contrast, the strong anti-realist believes that

(AR_1) an explanation of the meaning of a statement *S* consists in giving justification-conditions: the conditions for *S* that justify (provide adequate evidence or warrant for) asserting *S*;

(AR_2) an explanation of the meaning of a predicate *P* consists in giving ascription-conditions: the conditions for *P* that justify (provide adequate evidence or warrant for) ascribing *P* to an individual *x*.

On the anti-realist's view the understanding of a statement consists in knowing what has to be the case for the statement to be justifiably asserted. The justification-condition is a condition that can be recognized as obtaining when it does obtain.

The anti-realist claims that the way we learn the meaning of expressions in our language, and in general the way we use our language, cannot support the realist's notion of a language-independent conception of a statement's truth. As Michael Dummett (1969: 244) has put it:

> In the very nature of the case, we could not possibly have come to understand what it would be for the statement to be true independently of that which we have learned to treat as establishing its truth: there simply was no means by which we could be shown this.

The only notion of truth that a speaker could have acquired from his training in the use of a statement is a notion that coincides with the justifiability of assertions of the statement.

When a philosopher claims that the meaning of a statement is determined by its use, it is because he takes the meaning to consist in the role it plays in communication beween members of a speech community, and that each members's grasp of the meaning of an expression in his community's language consists in his capacity to use the expression in communication. In learning statements in his language, he learns what counts for and against them. This sentiment is expressed in a remark of Wittengstein's (1970: 66–66e):

> *Zettel* ¶437. The causes of our belief in a proposition are indeed irrelevant to the question of what we believe. Not so the grounds, which are grammatically related to the proposition, and tell us what proposition it is.

Our speaker also learns what place the statement has in sequences of inferences and what (perlocutionary) effects its utterance has on his addressees.

On the realist's view the audience's understanding of the sentence 'The *F* is not *G*' consists in knowing that associated with the surface structure are two distinct propositions or truth-conditions, and what those distinct truth-conditions are. This is how the realist explains the reported linguistic intuitions of speakers. When informants are presented with the sentence, they quickly recognize that they can understand the sentence in more than one way, and they recognize what those ways are. They report, "It might mean —, or it might mean . . .". And that means, alleges the realist, that the informant has grasped the two senses of the sentence. ('The *F* is not *G*' has three realist readings if we separate the presuppositions of existence and uniqueness. For the sake of exposition I have not done so here.)

These two senses, the realist continues, are mutually exclusive and jointly exhaustive alternatives. In any speech context an utterance of the sentence either has one sense, or it has the other. There is no third possibility. Here is a law of excluded middle. That, says the realist, is just what the informant meant when he said, "It might mean —, or it might mean . . .". He is in a position to know that there is an either/or choice just because he knows that the sentence is ambiguous. (The logic of this argument is, of course, independent of the number of readings.)

The realist is prepared to go beyond the testimony of the informant. Just as the principle of bivalence (that a statement is either true or false) stands behind the law of excluded middle ($\Vdash A \vee - A$), so also the realist believes that behind the linguistic excluded middle just mentioned stands the claim that the sentence *has* to mean — or . . ., in the same sense in which a statement *has* to be true or false. This is not what the informant reported in reporting his linguistic intuitions. It is a consequence of the realist's theory: how a sentence is used is determined by what sense it has.

Of course, in any speech context in which our speaker takes the *utterance* to have the one meaning he will usually say that *if* it means the one it could not also mean the other. But that is not enough to show that in any context an *utterance* of the sentence does or must mean precisely one or the other. And it certainly does not mean that the *sentence* does or must mean precisely one or the other. In fact, I claim that there is nothing in the words themselves, independently of any collateral information in the context of utterance, by virtue of which an ordinary speaker of English could claim that the sentence *had* to mean one or the other. I believe that our speaker could give no sense to that 'must'. Indeed, I believe that when the sentence is examined in isolation from normal contexts the ordinary speaker's intuitions about it oppose the realist view. His intuitions are describable by saying that the sentence does *not* have to mean the one or the other. When I utter *I am going to the game*, does the *sentence* 'I am going to the game' have to mean precisely that I'm going to the football game, or that I'm going to the soccer game, or that I'm going to the tennis game, etc? Surely not; nor is it synonymous with the disjunction of all those alternatives (as we saw in Chapter 2, Section 3).

What, then, is the explanation for the linguistic intuitions? I believe that when an ordinary speaker, confronting the sentence outside normal contexts, is prepared to say that the sentence does not *have* to mean the one understanding or the other, this is because the sentence is *not* ambiguous between them. The sentence is sense-non-specific between them instead.

If it is indeed true that the sentence is sense-general, then it certainly follows from that sense-generality that the *sentence* will not have the one understanding, or the other, or even the disjunction in the manner expected by the realist in his ambiguity account. Sense-generality *explains* our intuitions.

The ambiguity account is appealing to a realist because it makes a realism easier to defend. A sentence with different senses, each sense identifiable with truth-conditions, poses no special problem. The little Fregean in each of us is appeased, and it is predictable that logicians would find this account of the linguistic phenomena especially seductive. For many years I certainly did. The account is wrong, I believe, and a logician gives way to it like Aristotle's incontinent man faced with overwhelming temptation.

If 'The *F* is not *G*' is ambiguous, what I know is at least two distinct truth-conditions that can be described in an elementary way, say by Russell's Theory of Definite Descriptions:

(L^{+}) $(\exists x)((\forall y)(Fy \equiv y = x) \; \& \; - Gx)$
The *F* is non-*G*

(L^-) $-(\exists x)((\forall y)(Fy \equiv y = x) \& Gx)$
It is not the case that the *F* is *G*

But, if the sentence is sense-general it is *semantically neutral with respect to the presupposition that there is a unique F.* In that case 'not' is not identical with any familiar, logical operator; nor can it be accommodated by any familiar scope-delimiting device. Since the reading of the sentence is not the disjunction of L^+ and L^-, the realist is hard pressed to *say* just what the truth-conditions of the *sentence* are, the knowledge of which he must possess if he understands the sentence at all.

The Truth Theorist or the Possible-World Semanticist needs a way to represent the sense, i.e. (on his conception of sense) the truth-conditions, of the sentence, Neither L^+, L^-, nor $L^+ \vee L^-$ will do. Since the sentence is not ambiguous, it does not have one sense that is represented by L^+ and another sense that is represented by L^-. So, how does the semantic realist represent the sense of the sentence? The difficulty is in the nature of the linguistic intuition to be captured. For, though the sentence has no more than one meaning it can be understood in more than one way. An utterance of 'The king of France is not wise' can be understood to assert that the king of France is such that he fails to be wise (an understanding that normally "presupposes" that the king exists). It can also be understood to deny the proposition that the king of France is wise (without the existential presupposition). If understanding the meaning of the sentence involves a pairing of surface and underlying structure, which underlying structure is involved? Since the sentence is not ambiguous, the different understandings of the sentence cannot be explained by appeal to two different semantic representations. The difficulty would have been equally acute for Chomskyan Syntacticians, Generative Semanticists, or philosophers who accepted the Davidson (1970)-Harman (1970) hypothesis that underlying structure is identical with logical form. (For an interesting exposition of the Harman-Davidson programme, see Lycan (1984).)

If the object of our quest is a logical form that is sense-non-specific between exclusion and choice negation, the situation is clear. There isn't any. Formal languages as we know them are not designed to represent this aspect of grammatical meaning. If underlying structures determine grammatical meaning, and sense-generality is a semantic property of English words and sentences, the Harman-Davidson hypothesis is false. (For these negative sentences, if there are underlying structures but no logical forms, the statement that identifies them is false, or, if you will, at least not true.) The related hypothesis of Generative Semantics that underlying structures are both syntactic and semantic representations faces similar difficulties if the underlying structure is, more or less, a well-formed formula in a standard formal language containing the usual negations.[3]

The problem might not be quite so difficult if the sentence were "vague" rather than sense-general. Formal semanticists have at least offered theories of vague expressions, e.g. Goguen (1969), Machina (1972), and Lewis (1972). But vagueness as philosophical and logical semanticists understand it differs from sense-generality (see Chapter 2, n. 19). There is no help from that quarter.

Not finding much solace in extensional logic, the realist can retreat

[3] Another possibility is a representation adapting to our own purposes H. Paul Grice's (1981) bracket notation. Let 'The *F* is not *G*' be represented by:

(L) $- [(\exists x)((\forall y)(Fy \equiv y = x)\ \&]\ Gx)$

In order to "interpret" this formula we add to the semantics a Bracket Wipe-Out Rule (BWR; I take this terminology from Katz and Langendoen (1976)) according to which:

(BWR) $X - [C] - Y \rightarrow X - C - Y$,

and a Bracket Rewrite Rule (BRR) according to which:

(BRR) $X - [C] - Y \rightarrow C - X - C - Y$.

The metavariables here range over sequences of primitive symbols, not over formulae or constituent phrases. Our problem is to represent *one* sense, here allegedly given by L, and accommodate *more than one* understanding of the sentence, viz. the predicate negation L^+ and the sentence negation L^-. If we analyse L as follows:

$X \equiv -\ ;\ [C] = [(\exists x)((\forall y)(Fy \equiv y = x)\ \&];\ Y = Gx)$

BWR produces L^-; BRR produces:

$(\exists x)((\forall y)(Fy \equiv y = x)\ \&\ - (\exists x)((\forall y)\ (Fy \equiv y = x)\ \&\ Gx)$

which is equivalent to L^+ (except for parentheses, which require another rule).

The syntax for the bracket rules is messy because the contents of the brackets are not constituent phrases or well-formed formulae. But, more to the point, even though the machinery produces two formulae from a single representation L it is not clear what the bracketed formula L is. Is it really a semantic representation of the sentence? Is it a sense? Does it have truth-conditions? (What are they?) What does the fact that we can get semantic representations of the familiar kind from an unfamiliar representation and rewrite rules actually mean? How does this mechanism distinguish ambiguity from sense-generality? That is, what are the conditions under which the rules BWR and BRR are applicable to L? Well, there are no good answers to these questions. So I shall leave Grice's bracket device aside.

to intensional logic. Thus, following a suggestion of Montague (1970), Stalnaker (1972), and Kaplan (1979), let the *meaning* or *character* of the sentence be represented by a function from contexts of utterance into propositions, *Meaning*: *Contexts* → *Propositions*.

This is tempting because it relies on realist assumptions and makes it possible to associate a sentence with different propositions without implying that the sentence is ambiguous. For example, if in 'The *F* is *G*' the singular term 'The *F*' is referentially ambiguous, referring on one occasion of utterance to one individual and on another occasion to another, we can reflect the change in propositions expressed without claiming that the *sentence* is ambiguous. To know the meaning of the sentence is to know the rule (the function) according to which the meaning and contexts of utterance determine what proposition the sentence expresses in each context. (If in context K_a, K_b, . . . the term 'the *F*' refers to *a*, *b*, . . ., then by one possible rule the proposition given the sentence in each context is '*a* is *G*', *b* is *G*', ...) Since understanding *propositions* consists in knowing their truth-conditions, understanding the *sentence* consists (in part) in knowing truth-conditions.

Unfortunately, this is no solution to the realist's problem. In the case of the pragmatic intension, the effect of sense-generality is clear. In *any* context, a sense-general sentence can express *any* of its propositions. The realist's intension is not even a well-defined function. In *any* context 'The *F* is not *G*' can express either negative proposition, in so far as it is in the power of *words* to express anything.

The problem for the realist is that he's looking for a single, well-defined proposition, a Fregean *Gedanke*, with precise truth-conditions. If the sentence were merely ambiguous, the individual senses would constitute well-defined propositions. But, if not, how shall the realist describe what the speaker knows? The extensionalist and intensionalist efforts that we have considered here have not been satisfactory.

But, worse than that, sense-generality and Davidson's (1967*b*) semantic realism are incompatible, i.e. the sense-generality of 'The *F* is not *G*' is inconsistent with a truth-conditions conception of sense. If we employ John McDowell's (1976: 50) characterization:

A theory of truth, serving as a theory of sense for a language, must show

how to derive, for each indicative sentence of the language, a theorem of the form 's is true if and only if p', where what replaces 'p' in each case is . . . a sentence giving the content of propositional acts which speakers of the language can intelligibly be regarded as performing, or potentially performing, with utterances of the sentence designated by what replaces 's'

sense-generality entails that we ought to have as theorems:

(i) 'The *F* is not *G*' is true if and only if L^-

and

(ii) 'The *F* is not *G*' is true if and only if L^+.

Such biconditionals are not both derivable; if they were, it would prove that $L^- \equiv L^+$, which is absurd. So, either McDowell's characterization is incorrect or it is impossible to accommodate the sense-generality of 'The *F* is not *G*' in a truth-theory-based theory of meaning.

One temptation is to try to capture sense-generality by

(iii) 'The *F* is not *G*' is true if and only if $L^- \vee L^+$.

But, applying McDowell's characterization, the imputation of the performance of a speech act whose propositional content is this logical disjunction is implausible for two reasons. First, the claim denies the existence of 'The *F* is not *G*' -speech-acts in which the propositional content is precisely one of: L^-, L^+. Second, $L^- \vee L^+$ is actually a defective representation, since it is logically equivalent to L^-, which is equally the propositional content of the utterance. (L^+ entails L^-.) This is obviously contrary to the intention of (iii). The logical disjunction must be abandoned as a semantic representation of 'The *F* is not *G*'.

The speaker of English knows, in the usual implicit sense, that in a *sentence* 'The *F* is not *G*' the *word* 'not' is sense-general. This knowledge is not captured by a truth-conditional theory of sense. What can the anti-realist say about the speaker's knowledge? In his view what the speaker knows is what recognizable circumstances count as justifying the assertion of 'The *F* is not *G*'. There are different possible *grounds* for the *correct employment* of the sentence 'The *F* is not *G*'. One ground consists in part of circumstances in which there is no unique *F*. Another ground consists in part of circumstances in which the *F* exists but is not *G*. This is the anti-realist's explanation of our informant's linguistic intuitions. When our informant said of the sentence that it might mean ——, or it might mean . . ., he was distinguishing different grounds for asserting the sense-general sentence. He was not distinguishing between truth-conditions associated with distinct readings of an ambiguous sentence. This anti-realist explanation seems to me better than anything the truth theorist or the possible-world semanticist can offer. The anti-realist can give an explanation of a speaker's knowledge of presuppositional phenomena in natural language, of intuitions about the use of negative sentences with singular terms, so-called "factive" verbs, adverbials, and perhaps quantifiers, that is consistent with the linguistic facts of sense-generality. But, as Sir Peter Strawson cautioned me, one should *not* conclude that the *only* explanation is an anti-realist explanation. With this, of course, I agree.

My primary aim is to show that sense-generality is demonstrably incompatible with standard forms of truth-theoretic theories of meaning and with the resources of possible-world semantics. The linguistic phenomenon of sense-generality demands a new perspective from which to view the relation between the truth-value of *propositions* and the meaning of the *sentence* or sentences used to express them, i.e. a stance from which to evaluate the realist's theory of meaning.

It is also a perspective from which to view George Lakoff's version of Generative Semantics, where logical forms determine the literal meaning of a sentence in a context. But what logical form is associated with a sentence can change with the context. Where Lakoff would, I assume, posit two derivations to account for the different understandings of the negative sentence in different contexts, I would posit only one. Since the negative sentence is semantically "scope"-non-specific, it may be understood as either a sentence negation or a predicate negation. But it is not ambiguous between these understandings; however it is understood in a given context, the *words* do not *have* to be understood that way. If we say that this unambiguous sentence may be used to express different propositions, which may differ in truth-value, these propositions are not properly construed as "meanings" of the sentence. The sentence has one meaning. It is part of understanding the meaning of 'The king of France is not wise' to know that it can be used to express both a weaker and a stronger negative proposition. If semantic representations represent meaning, they are NOT propositions or logical forms, though which propositions can be literally expressed by a sentence is delimited or constrained by its semantic representation. A semantic representation of a negative sentence is a much more "abstract" object than is a logical form or a set of possible worlds.

2. Do Differences in Reference Entail Differences of Meaning? The Evidence from Natural-Kind Terms[4]

Hilary Putnam's (1973) "Meaning and Reference" contains an interesting passage that is absent from his (1975) "The Meaning of 'Meaning'". The passage argues against the possibility of a case in which 'water' in English refers to H_2O on Earth, 'water' in Twin English refers to XYZ on Twin Earth, but the words have the same *meaning*. This would of course require the abandonment of the principle that differences of reference entail differences of meaning, which Putnam does not wish to abandon. Putnam's (1973: 71 n. 2) *reductio* argument is this:

> Suppose 'water' has the same meaning on Earth and on Twin Earth. Now, let the word 'water' become phonemically different on Twin Earth—say it becomes 'quaxel'. Presumably, this is not a change in meaning *per se*, on any view. So 'water' and 'quaxel' have the same meaning (although they refer to different liquids). But this is highly counter-intuitive. Why not say, then, that 'elm' in my idiolect has the same meaning as 'beech' in your idiolect, although they refer to different trees?

Indeed, why not? If *words* refer to objects, it is not by the grace of Plato alone. It is by co-operative linguistic behaviour underlain by the grammar of a language. In a fairly large English speech community any speaker knows that the words "refer to different trees", though he may not be able to distinguish beeches from elms nor even say very much at all about what they are like. Even where inter-subjective conventions of use obtain we have occasion to use the word 'meaning' to describe cases in which one person's idiolect employs 'elm' in just the way another person's idiolect employs 'beech'.

For example, in section 649 of *On Certainty* Wittgenstein (1969) writes:

[4] This section is extracted from Atlas (1980*b*), © 1980, by Basil Blackwell Publisher Ltd. For very similar but independent arguments see Keith Donnellan (1983).

I once said to someone—in English—that the shape of a certain branch was typical of the branch of an elm, which my companion denied. Then we came past some ashes, and I said "There, you see, here are the branches I was speaking about." To which he replied "But that's an ash"—and I said "I always meant ash when I said elm."

Wittgenstein's usage diverged from that of his friend, but because they inhabited a common world the divergence of Wittgenstein's usage indicated a divergence from his friend's language. Thus it is appropriate for Wittgenstein to say that when he *said* 'elm' he *meant* ash. This is not merely to say that his utterance of a sequence of phonemes 'elm' succeeded in referring to an ash. Let us suppose Wittgenstein typically talked this way about ashes. When he used the word 'elm' he meant what his friend would have meant in using the word 'ash'. We describe this by saying that in Wittgenstein's idiolect 'elm' has the same meaning as 'ash' in his friend's idiolect. It seems to me that this employment of the word 'meaning' is not out of place.

However plausible, these considerations do not confront Putnam's science-fiction story, so I turn to a variation of Putnam's example. Suppose that 'water' in Twin Earth dialect of English refers to D_2O, deuterium oxide, deuterium being an isotope of hydrogen, D_2O naturally occurring in (Earth) water as approximately 1 part in 6500. My previous sentence is an ordinary as well as scientific way of talking that shows that 'water' in Earth-English does not in fact refer only to H_2O, but to H_2O and any of its isotopic relatives. (I shall ignore the complexity introduced by the solubility of metallic salts.) In my science-fiction story we are supposing that on Twin Earth the counterpart of (Earth) water in rain, lakes, oceans, etc., is indeed just D_2O. On Earth we refer to pure samples of D_2O by 'heavy $water_e$'. We refer to pure samples of H_2O by '$water_e$'. We (scientists also) refer to naturally occurring mixtures by '$water_e$'. Aggregates that Twin Earth speakers call '$water_{te}$', Earth speakers call 'heavy $water_e$' when pure, and '$water_e$' if mixed with what Earth speakers also call '$water_e$'. In sum, on Twin Earth '$water_{te}$' denotes D_2O. On Earth '$water_e$' denotes X_2O, where 'X' ranges over Hydrogen and its isotopes. (Oxygen has isotopes, but that complexity will not change the logic of the argument and can be ignored.)

Putnam supposes that there is phonemic change on Twin Earth so that '$water_{te}$' in Twin English becomes 'quaxel'. '$Water_e$'

denotes X_2O; 'quaxel' denotes D_2O. Putnam's argument is then a *reductio*—Supposition: 'Water' and 'quaxel' have the same meaning. Ho, ho, ho, ho, ho! Conclusion: 'Water' and 'quaxel' do not have the same meaning. This argument makes one suspect that its underlying logical form is " 'water' and 'quaxel' do not have the same meaning. Therefore 'water' and 'quaxel' do not have the same meaning."

When in "The Meaning of 'Meaning'" Putnam treats an example like the one I have described, he remarks that "if H_2O and XYZ had both been plentiful on Earth, then . . . it would have been correct to say that there were *two kinds of 'water'*." And instead of saying that "the stuff on Twin Earth turned out not really to be water" we would have to say "it turned out to be the *XYZ kind of water*" (Putnam 1975: 160). D_2O is not exactly what Putnam has in mind here; he envisages lakes of H_2O and lakes of XYZ on Earth. But there is enough similarity to suggest that Putnam should agree that in my story there are at least two kinds of water. Since in fact we use the nomenclature of 'heavy water', Putnam's reasoning is confirmed. Heavy water is a kind of water. So far as I can see, what Putnam does not even ask in the case of kinds of water is whether '$water_e$' and '$water_{te}$' have the same meaning. What decides the meaning of 'water' on Earth and Twin Earth? Is it that the words denote aggregates belonging to distinct species of the genus water, or that the words denote aggregates of the genus water of whatever kind? If the former, then '$water_e$' and '$water_{te}$' have different meanings. If the latter, then '$water_e$' and '$water_{te}$' have the same meaning.

In fact, since heavy water is water, '$water_e$' denotes aggregates of water of whatever kind. *Ex hypothesi* '$water_{te}$' denotes only D_2O, but nothing will have stood in the way of the Twin Earth physicists discovering isotopes and realizing that they could in principle produce "light $water_{te}$". This suffices to show that '$water_{te}$' is also a term with generic meaning. '$Water_e$' and '$water_{te}$' (or equivalently, 'water' and 'quaxel') denote aggregates of different liquids, i.e. they differ in reference, but nonetheless '$water_e$' and '$water_{te}$' have the same (generic) meaning. Natural-kind terms are analogous (but only analogous—see Chapter 2, Section 0) to deictic terms, e.g. 'I', whose referents differ in different idiolects but whose meaning is constant; so the doctrine that differences in reference (extension) entail differences in meaning (intension) is incorrect. My argument will generalize, showing that other natural-kind words are sense-general, e.g. 'man' and 'herring-gull'. I conclude that difference in extension is *not*, *ipso facto*, a difference in meaning for natural-kind words. *One prop under Putnam's claim that meanings are not concepts is thereby removed.* It may also be worth remarking that another such prop, the Division of Linguistic Labour, the emphasis on which Putnam properly regards as an original feature of his discussion, was anticipated by Wittgenstein (1969) in his Division of Epistemic Labour (*On Certainty*, Sections 162, 621, *passim*) and Donnellan (1962).

Postscript

After doing what I thought was linguistic invention in the language of Twin Earth physicists, and hypothesizing that they would refer to H_2O by 'light $water_{tc}$', I was astonished, and pleased, to discover that the doubles of Twin Earth scientists back on Earth were already speaking my hypothetical Twin English. I have learned that nuclear-reactor engineers refer to reactors that use "ordinary water" to cool the reactor's uranium fuel—as contrasted with the original use of heavy water—by the phrase 'light water reactors'.

3. Utterance Inference and Sentence Generality: The Construction of Utterance-Meaning[5]

H. Paul Grice's (1975, 1978, 1981) description of a conversational inference is roughly as follows. For a speaker S and an addressee A, A knows that in uttering *P* S means to communicate *Q* because in the conversation S has said *P*; there is no reason to suppose that he is not observing the maxims (saying what he knows, being relevant, being informative, being clear) or at least the Co-operative Principle (make your conversational contribution such as is required, at that stage at which it occurs, by the accepted purpose or direction of the talk exchange in which you are engaged); S could not be observing the maxims unless he thought that *Q*; S knows (and knows that A knows that S knows) that A

[5] This section includes a reorganized, abbreviated, and revised version of material extracted from Atlas (1979: 270–9), © 1979, by Academic Press, Inc.

can see that the supposition that S thinks that *Q* is required; S has done nothing to stop A thinking that *Q*; therefore S intends A to think, or is at least willing to allow A to think, that *Q*; and so: S has implicated that *Q* (Grice 1975: 70).

Such an account imputes to the speaker/hearer a knowledge of the semantic representation of a sentence, a conversational competence to assess the co-operative fit between a given context and the semantic representation, and an inferential capacity to discover or construct an alternative proposition from the semantic representation and the context.

The claims for the explanatory power of Gricean principles of conversational inference rest upon a highly convincing if vague account of the relationship between the meaning of the implicans, its conversational role, and the resulting implicata. But, for words and sentences the theory posits meanings that are controversial, while seeming to assume that the adequacy of such posits of word/sentence meaning can never be tested directly, but—and this is a truism in the theory—only by their contribution to the speaker's utterance-meaning of words and sentences uttered in contexts of actual use. The theory's great success is its convincing explanation of how and what speakers are understood to mean when patently they do NOT mean what their words do. It would be perfectly cogent for the theory to claim, for some class of words or sentences, that NO speaker EVER means what his words mean. For example, positing the truth-functional material conditional as the meaning of the English *if . . . then* would yield a case very close to this theoretical extreme.[6] Also, the obverse case, the theory treats as "degenerate" any case where the speaker means precisely what his words mean.

The theory is arguably adequate as a theory of conversational inference; in particular as a theory of what a listener could infer from a speaker's sentence and its context if the listener assumed that the speaker's words such as 'not', 'or', 'and', and 'if . . . then' and the speaker's sentences were synonymous with $-$, $\vee$, &, and $\rightarrow$ and with logical forms.[7] Indeed, the account is enlightening.

[6] It is too often forgotten that Russell operated with a principle of logical economy: minimize logical content. This aspect of Russell's use of the material conditional for *if . . . then* is insufficiently appreciated. See Russell (1919*b*).

[7] It turns out that Grice's paradigm of reasoning from his Maxims will generate contradictory, First-Maxim-of-Quantity, scalar implicata from an assertion of a

But how much CAN this kind of theory tell us about the ACTUAL meanings of English expressions?

Suppose one could successfully sail, for example, from Dover to Alexandria, by using a theory of ocean navigation in which it was assumed that the earth is uniformly flat. One might admire the theory because much might be learned from such a theory about the nature of rational principles governing the activity of navigating. In particular, one might learn much by studying the inferences employed in getting from paths in a two-dimensional Euclidean plane to paths on a three-dimensional almost-Euclidean sphere. It would be perfectly cogent for the theory to claim, for some class of theoretically possible voyages, that NO sailor EVER makes them—for example, a voyage beyond the edge of the world. Moreover, the theory treats as "degenerate" any case where the actual path of the voyage and the theoretical path of the voyage are (to a very close approximation) the same—for example, Dover to Calais. These are not very interesting voyages, since the theory's principles of inference do no interesting work—as it were, *A*, therefore *A*.

Despite telling us much that is interesting and even true about how to get from *a* to a distant *b*, how much CAN this kind of theory tell us about the way the world ACTUALLY is in the vicinity of *a*? Basically, for any *a* that is a possible port of call the theory says it is very flat around *a*. Unfortunately, though our theory gets us from Anchorage to Seattle, we KNOW it is not very flat around Anchorage, and we did not do any sailing to find that out.

Now suppose, by using a theory of conversational inference in which it was assumed that some English words, among them adverbs and subordinating and co-ordinating conjunctions, were

logically strong, negative, scalar sentence, and so is inadequate in the most logically embarrassing way. For the diagnosis and a resolution of the contradiction, see Atlas (1984*a*). Atlas and Levinson (1981) and Horn (1984*b*) show that First-Maxim-of-Quantity implicata can be inconsistent with Informativeness (Atlas and Levinson 1981), or "R-based" (Horn 1984*b*), Second-Maxim-of-Quantity implicata. Gazdar (1979) shows the inconsistency between scalar and clausal First-Maxim-of-Quantity implicata. For discussion of the strong, negative, scalar sentences, see Atlas (1984*a*) and Horn (1985), Kempson (1986), and Levinson (1988). The construction of good theoretical resolutions of these inconsistencies is a major concern of current linguistic research. See the discussion in "Section 7.1.2 Pragmatics", in Newmeyer's (1986: 174–9) *Linguistic Theory in America*, and in Horn's (1988) "Pragmatic Theory", in *Linguistics: The Cambridge Survey, Volume 1: Linguistic Theory—Foundations* (Newmeyer 1988: i, 113–45).

respectively synonymous with the various logical connectives, one could successfully explain what a listener could take a speaker to mean. How much CAN this kind of theory tell us about the ACTUAL meanings of English words like 'not', 'only if', and 'every'? Though our theory of conversational inference gets us from a negative presuppositional sentence to a choice-negation understanding of an utterance of the sentence in the appropriate context, as we shall see, we KNOW that the 'not' in the negative sentence is neither ambiguous between an exclusion and a choice negation nor identical with either, and we did not do any implicating to find that out.

In short, the structure of Grice's theory is problematic in at least two respects: On the one hand, the theory allows the possibility that speakers never mean what they literally say, and on the other hand, when they do, the theory provides no explanation of it. Grice's theory claims, falsely, that the literal meaning of the English word 'not' is that of an ordinary logical connective.

In what follows I wish to suggest novel kinds of arguments in support of distinguishing semantic representations of sentences from Russellian logical forms. I shall suggest that a methodologically satisfying Radical Pragmatics demands an equally Radical Semantics.[8]

Taking a negative sentence in isolation competent speakers know that it has (at least) two uses or understandings. Independently of context, the understandings are phenomenologically of equal status, neither judged less a function of the meaning of the sentence than the other. But the account in a Gricean theory is "unequal", in that it is split between the semantics and the pragmatics, and the different understandings are of different theoretical status.

For example, Boër and Lycan (1976) argued that IF the negative sentence were the sentential negation it did not necessitate—and so the affirmative sentence 'The king of France is bald' did not semantically presuppose—that there is a king of France; and that IF the negative sentence were the predicate negation, which does necessitate that there is a king of France, the negative sentence is not THE DENIAL of the affirmative. In either case, therefore, there is no semantic presupposition.

[8] Compare the notion of "partially specified semantics" in Lakoff (1977). See also Lakoff (1987).

Their positive account was a Gricean account, in which the addressee asks himself how the speaker knows, since he is enjoined by Grice's maxims to say what he knows or justifiably believes, that the king of France is not bald, and in answer infers that the speaker's evidence is that the (extant, unique) king of France is non-bald, the predicate-negation reading. This reading entails the existence of a king, the "existential presupposition" (Boër and Lycan 1976: 48–9).

If the description of the speaker's knowledge is given by the sentence 'The king of France is not bald', and a possible meaning of that sentence is the sentential negation, then Boër and Lycan argued that from the sentential negation different inferences as to the speaker's evidence are possible. Hence, they were happy to acknowledge my observation that even the English sentence allegedly reporting this reading, viz. 'It is not true that the king of France is bald', has "presuppositions". (See Atlas (1974, 1975*b*, 1977*b*); Boër and Lycan (1976: 48–52); Horn (1978*a*, 1985); and Chapter 3, Section 1.) On the other hand, if the predicate negation is a meaning of 'The king of France is not bald', once the sentence is so understood the Gricean mechanism adds nothing to the understanding of the utterance. (For further discussion see Lycan (1984) and Atlas (1988*a*).)

Following Grice one could also argue that when a speaker says *The king of France is not bald* there is no reason to suppose that he is not observing the maxims. He could not be doing so unless he thought that the king of France were non-bald, the most informative (and relevant) literal claim he can make in uttering those words in any context *K*. He knows, and knows that I know he knows, that I can see that the supposition that he thinks that the (unique, extant) king of France is non-bald is required in the context about which we are reasoning. He has done nothing to stop me from understanding him this way. Therefore he intends me to think, by virtue of his uttering *The king of France is not bald*, that the unique, extant king of France is non-bald. That is, he implicates the predicate-negation understanding of the sentence. (See Atlas and Levinson 1981).

The Gricean view, without the sense-generality of negation, permits a negative utterance semantically interpreted as a sentence negation to implicate a predicate-negation proposition that entails the existential presupposition. (A negative utterance semantically interpreted as a predicate negation would straightforwardly entail it.) And of course there are contexts in which no implicature of the sentential negation is intended. Letting L^- stand for the sentential negation, L^+ for the predicate negation, the function PRAG for the Gricean inference, and K for kinds of context, we may abbreviate the Gricean claims by the formulae:

$$\text{PRAG}\ (K^*, L^-) = L^+,$$
$$\text{PRAG}\ (K^{**}, L^-) = L^-.$$

In the second case the pragmatics adds nothing to the semantic interpretation; in the first case it obviously does. The standard Gricean view permits this kind of asymmetry in the theory. The first case is paradigmatic; the second case is degenerate. Why should there be this difference?

No explanation is given for the degenerate case, where we may view the implicature produced in the paradigm case as cancelled by the contextual difference between K^* and K^{**}. One remark, appropriate in this theory but unilluminating, is that in K^{**} but not K^* the speaker means what his sentence literally means; another, equally circular, is that the speaker does not presuppose that there is a French king.

On Grice's (1975) own account the semantic representation or logical form L^- would express the literal meaning of the sentence, and the semantic representation would be a representation of the sentence's truth-conditions. On the Atlas (1975*a*/*b*) accounts the "semantic interpretation" was L^-, but the "semantic interpretation" did NOT express the SENTENCE'S truth-conditions. The SENTENCE did not have truth-conditions; only propositions expressed by it in contexts had truth-conditions. The "semantic interpretation", identified with the logical form L^-, was NOT taken to be a representation of the MEANING of the *sentence*; it was employed in a theory of what it is to UNDERSTAND UTTERANCES of the sentence, not in a theory of what the SENTENCE MEANS. It was *a hypothetical utterance-meaning* (a meaning of certain tokens, not the meaning of a speaker).

Gricean principles of inference constituted part of the theory of understanding, and inferential competence was taken to be part of one's capacity to understand an utterance of the sentence. One did not have knowledge of the truth-conditions of the utterance solely by virtue of knowing the meaning of the sentence uttered.

Knowledge of those truth-conditions was arrived at by inference from knowledge of meaning, of persons, and of the world.

The choice of L^- as the "semantic interpretation" of the negative *utterance* in my 1975 Gricean theory did not tell me what represented the meaning of the negative *sentence*. Indeed, in my 1975 Gricean theory it was not intended to. For, if a *semantic representation* is supposed to represent *directly* the linguistic facts about the meaning(s) of a sentence, the exclusion-negation logical form cannot be the semantic representation, since it fails to represent sense-generality. And, if my present arguments are cogent, the choice of L^- in Grice's form of the theory could not actually tell me, despite the pretence of doing so. There is only the appearance of inconsistency between the claim of a Gricean pragmatic theory that L^- is the "semantic interpretation" of utterances of 'The F is not G' and my claim that, because of the sense-generality of 'The F is not G', L^- cannot be a "semantic representation" of its sentence-meaning. The apparent inconsistency results from a mistaken conception of the role of L^- in Gricean theories. These theories do NOT provide an alternative semantic theory of negative, presuppositional SENTENCES. I shall now show that with the ordinary concept of a semantic representation of sentence-meaning a pragmatic theory that is more coherent than Grice's or the Griceans' is possible.

I have discussed the theoretical "asymmetry" in the standard pragmatic theory's account of the L^- and L^+ understandings of utterances of the sentence 'The F is not G'. A Gricean view combined with the representation of the sense-generality of negation remedies this theoretical "asymmetry" in the standard Gricean theory. Let $\mathfrak{L}$ stand for the sense-general semantic representation of 'The F is not G'. Then the pragmatic theory does theoretical work in BOTH cases:

$$\text{PRAG}(K^*, \mathfrak{L}) = L^+$$
$$\text{PRAG}(K^{**}, \mathfrak{L}) = L^-.$$

Understandings that phenomenologically are of equal status are now represented by the theory as being of equal status, produced in the same way by the same mechanism. The problem of explaining the degenerate case—that is, the case where PRAG $(K^{**}, L^-) = L^-$—simply vanishes.

It may be helpful to consider a parallel with phonology. The sentence 'The *F* is not *G*' in one context may be understood as L^- and in another as L^+. These understandings are related to the sense-general meaning $\mathfrak{L}$ of the sentence as ALLOPHONES are related to the ARCHI-PHONEME to which they belong (see Hyman 1975: 70–1).

Understanding an utterance is not simply knowing a logical form that the context SELECTS from the meanings of an ambiguous sentence. Understanding an utterance is knowing a proposition that in the context the hearer CONSTRUCTS from the meaning of a univocal, sense-general sentence. The general meaning of the sentence is made specific in the hearer's understanding of the utterance. This theory (Atlas 1979) implies that pragmatic inference assists in the construction of the logical form of (or proposition expressed by) an utterance. Gricean inference is not restricted to mapping propositions ("what is said") into propositions ("what is implicated"). It also maps Lockean, "abstract" sentence-meanings into propositions (utterance-meanings). For other defences of this theory of "pragmatic intrusion" into the truth-conditions of utterances, and its application to a wide range of linguistic data, see Kempson (1986), Levinson (1988), and Levinson (forthcoming).

For a comment on the relationship between Atlas's (1979) and Kempson's views, as well as further development of this theory, see Kempson (1988: 141 n. 2). In response to Kempson (1986), Horn (1989: 433) writes:

> Kempson 1986 maintains that while scalar predications are not ambiguous either lexically or at the level of semantic representation, they are ambiguous *propositionally*, at the level of enriched logical form. For Kempson, utterance interpretation is radically underspecified by linguistic meaning; pragmatic principles—including the familiar Gricean implicata—may (*contra* Grice) influence propositional content and hence help determine truth conditions. If she is right (and see Atlas 1979, Carston 1985*a*/*b* and Sperber and Wilson 1986 for parallel arguments), no straightforward distinction between what is *implicated* and what is *said* (as defended or assumed by Grice, Gazdar, Karttunen and Peters, and of course Horn) will survive.

It is, in fact, an immediate consequence of the views of Atlas (1979), Kempson's views, and the views of this section that *there is no straightforward distinction between "what is implicated" and "what is said"!* Matters of indexicality aside, pragmatic inference is constitutive of both.

The opposing Grice view is dubbed 'the semantic autonomy thesis' by Horn, in his "Pragmatics, Implicature, and Presupposition", W. Bright (ed.), *Oxford International Encyclopedia of Linguistics*, to appear, where he writes:

> Some potential problems for the semantic autonomy thesis are sketched by Gazdar (1979: 164–8), and recent studies by Sperber and Wilson (1986), Kempson (1986), and their colleagues, extending an argument due originally to Jay Atlas (see his paper in Oh and Dinneen 1979, pp. 275–9), have cast doubt on the tenability of the original Gricean schema. Evidence from various sources indicates that the grammatically and lexically assigned meaning of an expression radically underspecifies its truth-conditional content, and that the same considerations that enter into the post-hoc calculation of conveyed meaning also help determine propositional form.

Another analogy may be of some interest. When one makes a measurement on a Quantum-Mechanical system whose state function is $\Psi_t(x)$, the measuring process "develops" a value in the measurement of the observable property $\hat{O}$ by forcing the wave function to become one of the eigenfunctions $o_i(x)$, with the result that the measurement will produce the associated eigenvalue O_i. In advance, all we know is that the probability of this value being the value that we measure is calculable from the state of the system $\Psi_t(x)$ and the eigenfunction $o_i(x)$ together. O_i is not a value the observable property $\hat{O}$ "has"; it is obtained by measuring $\hat{O}$.

However, the state of the system is NOT determined by the result of any measurement that might be performed on it; it is determined by the Ψ function, and that is not a directly observable entity.

Likewise, a sentence-meaning $\mathfrak{L}$ coincides with neither an utterance-meaning L^+ nor an utterance-meaning L^-. The context "develops" a value for the utterance-meaning of the sentence. In a particular context that value may be the proposition L^+. As in the Quantum-Mechanical case, it is not even correct to say that the utterance-meaning of the sentence in the context IS the proposition (or, thinking of utterance-meaning as a function of utterance-context and sentence-meaning, "has" that proposition as its value). The very act of understanding ("measuring") the utterance DEVELOPS the proposition. It is not strictly correct to speak of the proposition expressed by the utterance; strictly, one should speak of the proposition understood by an addressee to have been expressed in an utterance by a speaker. The act of understanding "develops" a proposition by forcing the general sentence-meaning to be "developed" as a specific utterance-meaning (in the context in which the sentence is uttered), whose associated content is then, and only then, a proposition.

The wave function $\Psi_t(x)$ is "physically real", but it is not directly observed. Likewise, the semantic representation of the sentence, viz. $\mathfrak{L}$, is "psychologically real" (in linguistic competence), but it is not the entity directly meant (in performance) by the speaker.

There has been a belated recognition by Jerry Fodor (1983*b*: 90) that the "language-input system" may do *less* than specify, "for any utterance in its domain, its linguistic and maybe its logical [syntactic] form". Fodor (1983*b:* 135 n. 29) writes:

> Hilary Putnam has the following poser. Lincoln said, "You can fool all of the people some of the time." Did he mean *there is a time at which you can fool all of the people* or did he mean *for each person there is a time at which you can fool him*? Putnam thinks that Lincoln's intentions may have been *indeterminate* as between these readings. This could, of course, be true only if the specification of quantifier scope is not mandatory in the internal representation of one's intended utterances. And *that* could be true only if such representations *do not* specify the logical [syntactic] form of the intended utterance. To put it another way, on Putnam's view, the internal representation of "You can fool all of the people some of the time" would be something like "You can fool all of the people some of the time," this latter being a *univocal* formula which happens to have disjoint truth conditions. Whether Putnam is right about all this remains to be seen; but if he is, then perhaps the specifically *linguistic* processes in the production/perception of speech deploy representations that are *shallower* than logical [syntactic] forms.

When I asked Paul Benacerraf about Putnam's views he reported to me that Putnam did not remember posing this problem to Fodor. So I shall refer to the possibly fictional philosopher of this passage by 'Putnam*'.

My views are different from Putnam*'s, though there is a superficial resemblance. Putnam* is making a claim about a *mental* representation of an *utterance* in a speaker's mind. The analogous claim on my view, if the quantifers were scope-non-specific in the *semantic* representation of the *sentence*, would be, *pace* Putnam*, that there are no distinct "readings" for the speaker's intentions to be indeterminate between; the semantic representation itself is indeterminate between *there is a time at which you can fool all of the people* and *for each person there is a time at which you can fool him*. Paralleling Fodor, I would have to say: on Atlas's view the *semantic* representation of the *sentence* 'You can fool all of the people some of the time' would be something like 'You can fool all of the people some of the time', this latter being a *univocal* formula which happens to have disjoint [*not* disjunctive] truth-conditions. On my view the indeterminacy goes *deep*.

What is the relationship between semantic and pragmatic properties of sentences within a unified theory of meaning? In a theory of the kind that I have sketched Grice's pragmatics without sense-generality is blind, and sense-generality without Grice's pragmatics is empty. It is the mind's synthesis of the semantic and the pragmatic in understanding what is said that creates the deep and fascinating difficulties in the study of language.

Appendix
Metaphysical Ambiguity: Is 'Exists' Ambiguous?

Hardly anything seems more basic in our thought about the world than our distinguishing things as kinds: categorizing. Reflection on this cognitive activity fills pages of Aristotelian texts, and in our own late-nineteenth-century and early-twentieth-century logical tradition, as in the work of Bertrand Russell or in the later work of Gilbert Ryle, the ontological intuition that there are KINDS of things motivates the logical-linguistic intuition that there are "types" of words. But the types are not merely syntactic types, like "parts of speech", they are semantic types. Indeed, one "word" might *ambiguously* exemplify more than one logical type. Much logical and philosophical effort has gone into developing theories of such types. As profound as such theories are, my interest in this book lies in the notion of ambiguity, not, as would be traditional, in the notion of type or the notion of word. For, astonishingly enough, philosophers have given far less attention to ambiguity than its central employment in their reasoning would seem to require. To illustrate this employment I shall review briefly some specific history of twentieth-century English and American philosophy. I shall put historically familiar arguments in what I hope will prove to be a light both unfamiliar and illuminating to the reader. I shall begin with some of Bertrand Russell's comments on logical types and follow with a discussion of Gilbert Ryle. They are both concerned with whether 'exists' or 'there are' is ambiguous, being of more than one logical type. Then I shall review criticism of the ambiguity thesis by Morton G. White (1956/63) and find further grounds to support his criticism of Ryle and Russell.

Bertrand Russell (1924: 332) once wrote, with his cusomary insouciance, that

> Socrates and Aristotle are of the same type, because 'Socrates was a philosopher' and 'Aristotle was a philosopher' are both facts; Socrates and Caligula are of the same type, because 'Socrates was a philosopher' and 'Caligula was not a philosopher' are both facts. To love and to kill are of the same type, because 'Plato loved Socrates' and 'Plato did not kill Socrates' are both facts.

In 1924, prior to Tarski's (1936) work on the concept of truth in formalized languages, Russell tended to slip back and forth between object-language and metalanguage formulations, between types of individuals, attributes, relations, propositions and types of expressions naming or expressing individuals, attributes, relations, propositions.[1] In the following characterization of types

(T) A and B are of the same logical type if, and only if, given any fact of which A is a constituent, there is a corresponding fact which has B as a constituent, which either results by substituting B for A, or is the negation of what so results.

Russell (1924/56: 332–3) has characterized types of entities rather than types of expressions, since he regarded expressions as classes of sequences of noises or shapes and so all of one type. If one says "'Russell' means Russell" and "'Continuity' means continuity", where Russell is a different type of entity from continuity, the semantic facts just described, constituted as they are of different types of entity, are either different types of fact or are the same type of fact. Russell (1924/56: 334) had suggested that 'fact' itself was type-ambiguous. If facts are type-ambiguous, the definition (T) Russell has given for 'same type' is ill-defined. So let us assume for the moment that 'fact' is univocal.

Russell (1924/56: 332–3) assumed that it followed from his definition of 'same type' that

> when two words have meanings of different types, the relations of the words to what they mean are of different types; that is to say, there is not one relation of meaning between words and what they stand for, but as many relations of meaning, each of a different type, as there are logical types among the objects for which there are words.

In general, for relational statements $R(a,b)$ and $R'(a',b')$, if Type(a) = Type(a') but Type(b) $\neq$ Type(b'), then Type(R) $\neq$ Type(R').

Russell (1924/56: 336–7) took this principle even further. He wrote:

> I do not believe that there are complexes or unities in the same sense in which there are simples. I did believe this when I wrote *The Principles of Mathematics* (1903), but, on account of the doctrine of types, I have since abandoned this view. To speak loosely, I regard simples and complexes as always of different types. That is to say, the statements 'there are simples' and 'there are complexes' use the words 'there are' in different senses. But if I use the words 'there are' in the sense which they have in the statement 'there are simples', then the form of words 'there are not complexes' is neither true nor false, but meaningless.

In short, if Type(e) $\neq$ Type(e'), and both $A(e)$ and $A(e')$ are significant, then the sense of A associated with e is not the sense of A associated with e', and $^*A_e(e')$.[2]

[1] Nevertheless Russell (1918) in "The Philosophy of Logical Atomism" showed that he understood perfectly well the logical importance of the distinction between talk about objects and talk about talk.

[2] The asterisk '*' prefixed to a sentence or formula indicates that the sentence is ungrammatical or that the formula is ill-formed.

Max Black (1946: 229–55) raised the question, What if the relation in question is "*x* is the same type as *y*"? Suppose, Black argued, that *a* and *b* are of the same type, while *a* and *c* are of different types. Then the facts are:

(1) *b* is the same type as *a*.
(2) *c* is not the same type as *a*.

Since (2) is the negation of the result of substituting '*c*' for '*b*' in (1), Black asserted (falsely I believe) that, according to Russell's definition (T), *b* and *c* must be of the same logical type. By hypothesis they are of different types. Black argues that *a*, *b*, *c* must be the same type, whatever the character of *a*, *b*, or *c*. Since the content of the doctrine of types, as Russell (1924/56: 334) described it, was that "a word or symbol may form part of a significant proposition, and in this sense have meaning, without being always able to be substituted for another word or symbol in the same or some other proposition without producing nonsense", the conclusion of Black's argument (on the assumption that substituting expressions of the same type in a sentence preserves the significance of the sentence) denies the logical difference between *a*, *b*, and *c* that it was the point of the theory of types to assert.

Black pointed out that if Russell would reformulate his doctrine as one about types of expression rather than types of object the alleged difficulty could be avoided. Black argued as follows: If "*A*" and "*B*" are expressions of the same type, while "*A*" and "*C*" are expressions of different type, then statements about those expressions will take the form:

(1′) "*B*" is the same type as "*A*".
(2′) "*C*" is not the same type as "*A*".

If Russell's definition of type is now of the form:

("T") "*B*" and "*C*" are of the same type if and only if given any true sentence of which "*B*" is a constituent there is a corresponding true or false sentence which has "*C*" as a constituent, which results by substituting "*C*" for "*B*"

Black (falsely) concluded that "*B*" and "*C*" in (1′) and (2′) are of the same type. So "*A*", "*B*", and "*C*" must be of the same type. Black observes that names for expressions of different types are of the same type, so no absurd consequence follows.

In his reply to Black's criticism, Russell (1946: 691) accepted Black's (incorrectly argued for) strictures on applying 'type' to expressions rather than to individuals, attributes, or relations. He, correctly, did not accept a further criticism that Black offered of the type theory so applied. Black (1946: 237–8) had argued that the truth of

(1″) *I am thinking about Russell.*
(2″) *I am thinking about continuity.*

showed according to Russell that the expressions 'Russell' and 'continuity' were of the same logical type. This, Black thought, resurrected the original difficulty, undermining the very distinction it was the point of type theory to support. So Black suggested that Russell's characterization be changed to one for DIFFERENCE of types of expression:

> "*A*" and "*B*" are of different type if and only if there is some significant (i.e. true or false) sentence of which "*A*" is a constituent, whose syntactic transform under substitution of "*B*" for "*A*" becomes nonsense (i.e. neither true nor false).

Black further stipulated a condition of Russell's (1924/56: 332–3) own that if Type $(x) \neq$ Type(y), and $A(x)$ and $A(y)$ are both significant, then Type$(A_x) \neq$ Type(A_y). These conditions would show, first, that 'Russell' and 'continuity' are not of the same type, and, second, that 'thinking about' in 'I am thinking about Russell' and in 'I am thinking about continuity' are of distinct types.

It is historically interesting that Russell accepted Black's first criticism so readily. Although it raises an important question of typing relations that include the same-type relation, especially when the type hierarchy consists of individuals, attributes, and relations as opposed to the corresponding expressions, Black's argument does not correctly show that by Russell's definition of 'same type' *a*, *b*, and *c* in the first argument above must be of the same logical type. Russell's definition requires that for EVERY true sentence containing '*b*' its syntactic transform under the substitution operation of '*c*' for '*b* ' is significant, i.e. either true or false. Black's argument showed only that for SOME true sentence containing '*b*' its transform under substitution of '*c*' for '*b*' yields a false sentence, viz. '*c* is the same type as *a*'. That is not sufficient, according to Russell's definition, to show that *b* and *c* are the same type. Likewise, Black's second criticism, that Russell's condition showed that 'Russell' and 'continuity' were of the same logical type, is incorrect.[3] Since Russell's

[3] Black's example does not meet Russell's condition (T) because Black does not consider EVERY context containing '*b*'. Black is reading Russell's 'any fact' in (T) as 'some fact'. I am reading it as 'every fact', as do Kneale and Kneale (1962: 658). Black's reading of 'any' as 'some' seems to me not in the spirit of Russell's theory, but Russell himself in 1944 does accept Black's criticism. That gives me pause. Nevertheless, in the light of what Russell (1924/56: 332–3) has said about the typing of relations in "Logical Atomism" (1924), it seems to me that Russell should say to Black that '*x* is the same type as *y*' is a type-ambiguous expression. If '*b*' and '*a*' name entities of one type and '*c*' names an entity of another type, 'same type' in '*a* is the same type as *a*' does not mean the same as 'same type' in '*c* is the same type as *c*'. So Russell should have rejected Black's argument as equivocal. Russell should say that (2) is NOT the negation of (1) under the substitution of '*c*' for '*b*'. Moreover, the following argument is just like Black's (where $b = a$):

(1′) *a* is the same type as *a*.
(2) *c* is not the same type as *a*.

So, Black would conclude, on his version of Russell's (T):

formulation entails Black's formulation for *difference* of type, and, on the assumption—acceptable to Russell (1924/56: 334)—that for any sentence there is a negative sentence such that the former is true (false) if and only if the latter is false (true), Black's formulation entails Russell's for difference of type, there is nothing to choose between them. Except for the basic Tarskian distinction between object-language and metalanguage definitions of types, Black's emendations diverge not at all from Russell's original views. In his reply to Black, Russell (1946: 692) reformulates his view as follows (and in doing so shows that he is not as confused about the distinction between the use and mention of expressions as some have thought; at least he is clear about it by 1944):

> Words, in themselves, are all of the same type; they are classes of similar series of shapes or noises. They acquire their type-status through the syntactical rules to which they are subject. When I say that 'Socrates' and 'mankind' are of different types, I mean neither the words as physical occurrences, nor what they mean—for I should say that 'mankind' means nothing, though it can occur in significant sentences. Difference of type means difference of syntactical function. Two words of different types can occur in inverted commas in such a way that either can replace the other, but cannot replace each other when the inverted commas are absent.

It has struck more than one philosopher, e.g. Kneale and Kneale (1962: 671) and J. Passmore (1970: 131–2), that the criterion for difference of type is so fine-grained that hardly any two words will be the same type. A difference of two word-types would then be no different from a difference of two word-classes. The analogous criterion for the synonymy of two expressions was discussed by B. Mates (1952) and N. Goodman (1949): substitutivity *salva veritate* in every verbal context. If this condition is taken as necessary for synonymy, it is quite plausible to conclude that no expression can be synonymous with any other. Likewise, if substitutivity *salva significatione* in every verbal context is necessary for type-identity, it is quite plausible to conclude that no expression can be type-identical with another.

Yet we do not want to lose sight of the basic logical distinction Russell wanted to draw between 'Plato' and 'wisdom', that Frege (1884/1950) wanted to draw between 'wise' and 'one', and, on some conceptions of types as categories, that Aristotle wished to draw between processes and states, or that Ryle (1949) wished to draw between occurrent mental

a is the same type as *c*.

So, from the truism (1′) and the assumption (2), Russell's criterion allegedly implies the negation of (2). No sober-minded type-theorist should accept a criterion for same type such that the conjunction of a truism, an assumption, and the criterion entails the negation of the assumption. As before, Russell should say that (2) is not the negation of (1′) under the substitution of '*c*' for '*a*'.

episodes and dispositional mental states.[4] Frege's example, and a Russellian variant, perhaps point us in the right direction. Frege (1884/1950: 40e–41e) remarks that if 'Solon was wise' is true and 'Thales was wise' is true, then 'Solon and Thales were wise' is true. But 'Solon was one' is true, and 'Thales was one' is true, yet 'Solon and Thales were one' is not true. If being one were a property of Solon and Thales respectively, the argument would be valid; so Frege concludes that being one and being wise are not properties of the same type. Russell (1918/56: 233) makes a similar point when he writes,

> When you say 'Unicorns exist', you are not saying anything about any individual things, and the same applies when you say 'Men exist'. If you say 'men exist, and Socrates is a man, therefore Socrates exists', that is exactly the same sort of fallacy as it would be if you said 'Men are numerous, Socrates is a man, therefore Socrates is numerous', because existence is a predicate of a propositional function, or derivatively of a class.

The distinctions between 'Plato', 'wise', and 'one' or 'exists' are ones of logical type; they are required to explain formally the pre-theoretic distinctions we make between acceptable and unacceptable deductive arguments. Our formal theory of valid argument relies on distinctions among individuals, properties of individuals, and properties of properties, as well as among these and two-and-more-place relations.

Rules governing the possible grammatical relations among expressions of different types have also provided one way to evade the Russellian contradiction derivable from supposing there is such a property as the property of not exemplifying itself, i.e. that there is an *F* such that *not F(F)*.[5] Rather than merely demonstrate the logical impossibility of there being such a property, by deriving the contradiction, we redescribe the logically impossible as the logically "meaningless" according to rules governing the grammatical relations among the types of the symbols employed in the description of the property.

We introduce type distinctions among symbols to explain by formal means the cases of valid and invalid argument. We discover that some combinations of symbols lead to contradictory or otherwise absurd consequences. In order to avoid non-obvious and unrecognized contradictions that may lurk, we adopt grammatical rules that eliminate as

[4] Aristotle actually held a type-theoretic-like distinction between 'false' and 'meaningless'. See *Topics* 109a: 26–34.

[5] The property of non-self-exemplification is either self-exemplifying or non-self-exemplifying. If the latter, then the property IS self-exemplifying; if the former, then the property is NON-self-exemplifying. Thus the property is non-self-exemplifying if and only if it is self-exemplifying, i.e. [not F (F)] iff F (F), which is impossible. So, there is no such property, a property that is defined to be untrue of itself. The restrictions of type-theory make the description of such a property "meaningless", viz. ungrammatical.

logically "meaningless" the (logical) absurdity-breeding combinations of symbols that we have so far discovered. Such a strategy does not specify what manner or what degree of absurdity an inconsistency has to have before it is sensible to declare the source of the inconsistency a "meaningless" expression. Russell's strategy is far more radical than our long familiarity with it, and with analogous ones by Ryle and others, permits it to appear to us. An inconsistent statement, even of the most explicit kind, is not meaningless in other than a special Russellian sense. Something more than contradiction must be involved if 'meaningless' is aptly to apply. If we have any systematic criteria variously to distinguish among the unacceptable deductive arguments, they will be linguistic in character. Frege's example 'Solon and Thales were one' enthymemetically implies '2 = 1', and so is a straightforward contradiction arising from a mistaken analysis of the former sentence, but Russell's example 'Socrates is numerous' is in itself a semantically anomalous English sentence. So Russell (1919*a*/56: 233) was wrong to say that the 'exist' argument committed EXACTLY the same fallacy as the 'numerous' argument. Ordinary linguistic anomaly was not much emphasized by Frege and Russell, but as philosophy in the Russellian style took a linguistic turn even correct observations from grammar could, as Ryle (1949: 152) observed, father epistemological error on metaphysical innocence:

> It has long been realised that verbs like 'know', 'discover', 'solve', 'prove', 'perceive', 'see' and 'observe' (at least in certain standard uses of 'observe') are in an important way INCAPABLE OF BEING QUALIFIED BY ADVERBS like 'erroneously' and 'incorrectly'. Automatically construing these and kindred verbs as standing for special kinds of operations or experiences, some epistemologists have felt themselves obliged to postulate that people possess certain special inquiry procedures in following which they are subject to no risk of error. They need not, indeed they cannot, execute them carefully, for they provide no scope for care. The LOGICAL IMPOSSIBILITY of a discovery being fruitless, or of a proof being invalid, has been misconstrued as a quasi-causal impossibility of going astray. If only the proper road were followed, or if only the proper faculty were given its head, incorrigible observations or self-evident intuitions could not help ensuing. So men are sometimes infallible.

This is a case in which an awareness of semantic inconsistency in 'prove erroneously', 'know incorrectly', etc. suggests, if knowing is a doing, that there are incorrigible doings. The oddity of 'false knowledge' is evidence for the thesis that, necessarily, if McX knows that *P* then it is true that *P*. It is, as Ryle notes, a misconstrual to take this as causally necessary connections between certain facts and certain mental acts. As Ryle (1949: 151) remarked:

> Epistemologists have sometimes confessed to finding the supposed cognitive activities of seeing, hearing, and inferring oddly elusive. If I descry a hawk, I find the hawk, but I do not find my seeing of the hawk. My seeing of the hawk seems to be a queerly transparent sort of process, transparent in that while a hawk is

detected, nothing else is detected answering to the verb in 'see a hawk'. But the mystery dissolves when we realise that 'see', 'descry', and 'find' are not process words, experience words or activity words. They do not stand for perplexingly undetectable actions or reactions, any more than 'win' stands for a perplexingly undetectable bit of running or 'unlock' for an unreported bit of key-turning. The reason why I cannot catch myself seeing or deducing is that these verbs are of the WRONG TYPE TO COMPLETE THE PHRASE 'catch myself . . .'. The questions 'What are you doing? and 'What was he undergoing?' CANNOT BE ANSWERED by 'seeing', 'concluding'. . . .

In another vein Ryle (1949: 116) remarked, "The verbs 'know', 'possess' and 'aspire' do not behave like the verbs 'run', 'wake up' or 'tingle'; we CANNOT SAY [*without anomaly*] 'He knew so and so for two minutes, then stopped and started again after a breather',[6] 'He gradually aspired to be a bishop', or 'He is now engaged in possessing a bicycle'." From these linguistic anomalies—they are not merely inconsistent—Ryle concludes that knowing is not a process. In addition, he held that knowing is a disposition. But his evidence for this positive thesis was a dubious, intuitive synonymy claim, viz. that "to SAY that a person knows something . . . IS [*literally*] . . . TO SAY . . . that he is able to do certain things, when the need arises". It is, I think, clear that the synonymy claim is incredible. On the face of it, a speaker of English would not agree that 'McX knows something' MEANS 'McX is able to do certain things, when the need arises'.[7] Nevertheless, Ryle makes effective use of linguistic evidence from linguistic anomaly to scout the "category mistake" of confusing states with processes, an Aristotelian distinction of logical type.

His attack on the Cartesian doctrine that "every human being has both a body and a mind", which asserted that there are mental causes of bodily movement as well as physical causes of bodily movement, was an oblique one. His strategy was to reject a semantic assumption implicit in the doctrine. That assumption was that 'mind' and 'matter' are of the same logical type; hence it would make sense to oppose them, e.g. in 'mind versus matter', and to declare a winner, e.g. in 'a triumph of mind over matter'. In an exact parallel with Russell's (1924/56: 336–7) view, Ryle (1949: 22) held that "the phrase 'there occur mental processes' does not mean the same sort of thing as 'there occur physical processes', and, therefore, that it MAKES NO SENSE to conjoin ['there occur physical and mental processes'] or disjoin ['there occur physical or mental processes'] the two". Again like Russell, Ryle (1949: 23) claimed:

It is perfectly proper to say, in one logical tone of voice, that there exist minds, and to say, in another logical tone of voice, that there exist bodies. But these expressions do not indicate two different species of existence, for 'existence' is not a generic word like 'coloured' or 'sexed'. They indicate two different senses of 'exist',

[6] I might say this of an amnesiac, of course.
[7] Or Ryle is equivocating on 'say' in these passages.

somewhat as 'rising' has different senses in 'the tide is rising', 'hopes are rising', and 'the average age of death is rising'. A man would be thought to be making a poor joke who said that three things are now rising, namely the tide, hopes, and the average age of death. It would be just as good or bad a joke to say that there exist prime numbers and Wednesdays and public opinions and navies; or that there exist both minds and bodies.

Ryle distinguishes between the existences of mathematical entities (e.g. averages), mental entities, (e.g. hopes), and physical entities (e.g. tides) by distinguishing between three senses of the word 'exists'.

Morton G. White (1956/63: 65–6) offered the first systematic objections to this claim, when he observed that the ambiguity of 'exist' was just as doubtful as the view that 'are' in 'tides are higher', 'hopes are higher', and 'averages are higher' is ambiguous. Secondly, White observed that Ryle's thesis would be expressed by "There are two tokens of the phrase 'there are' in the sentence 'there are minds and there are bodies', and the two tokens together have two senses". Since tokens are physical marks on the page and senses are not, why wouldn't 'two' in 'two tokens' and 'two senses' itself have two senses? And, if so, Ryle could not say that the number of occurrences of 'there are' is two in the same sense that the number of senses of 'there are' is two (White 1956/63: 76–7). On his own grounds, Ryle's thesis could not be understood in the way in which, quite obviously, we do understand it. (In this objection White has done to Ryle what Black did to Russell.)

As a solution White suggested an analogy with terms related as genus terms are to their species terms, or as the word 'man' in a "general sense" is related to 'man' in a "special sense" (White 1956/63: 64). In the first sense 'man' multiply denotes human beings; in the second sense 'man' multiply denotes male human beings. As White (1956/63: 67) put it, "the notion expressed by ['man' in the first sense] is generic in the sense that it appears as a component of, or of some expansion of", the notion expressed by 'man' in the second sense. If the first sense of 'man' is that of 'human being', that sense is a component of the sense of 'male human being'. Thus 'exist' has a general sense that permits a philosopher like Ryle to invest 'exist' with a more specific sense, e.g. that of 'exists in space and time'. But this is to say that Ryle's phones [*exist*] are not the English word 'exist'. Whether the English word 'exist' has the specific sense is a genuine linguistic question. Perhaps, if it has this sense, it also has the sense of 'exists out of space and time'. How do we determine whether 'exist' is ambiguous (either homonymous or polysemous)?

Ryle (1949) offers a linguistic argument. He claims that the following conjunction is absurd:

(3) *She came home in a flood of tears and a sedan chair.*

It is a necessary, but not sufficient, condition on syntactic co-ordination by 'and' that the conjoined expressions have the same distribution of occurrence in other syntactic constructions in the language, and so be, by distributional criteria, the same linguistic "type". When this condition is violated, as in example (3), the result is known traditionally as zeugma. In this case from the alleged anomaly Ryle infers that 'in' is ambiguous.[8] Similarly, Ryle alleges that the linguistic absurdity of:

(4) *the tide, hopes, and the average age of death are rising.*

shows that 'rising' must have more than one sense. Finally, Ryle alleges that the following is absurd:

(5) *Minds and bodies exist.*

and that 'exist' must have more than one sense. Ryle seems to make two assumptions in his discussion. (i): the linguistic absurdity of the application of a meaningful expression *A* to a collection implies a difference in logical "category" among its members. In particular Ryle (1949: 22) asserted that if Type(*e*) = Type(*e'*), then *A(e and e')* is linguistically acceptable. (ii): a difference in logical category among the members implies a multiplicity of senses for any expression that is meaningfully applied to each of them (Ryle 1949: 23).[9] It follows from Ryle's two assumptions that the linguistic absurdity of an expression, in whose scope is a conjoined phrase, implies a multiplicity of sense of the expression, if the expression would be meaningful for each of the constituent phrases in its scope. If *A(e and e' and e")* is absurd, then *A(e)*, A(e'), A(e") has at least two occurrences of '*A*' in different senses if each of *A(e)*, *A(e')*, *A(e")* is not absurd.

Now, I do not hear the linguistic absurdity that Ryle expects me to hear in:

(5) *Minds and bodies exist.*

So, by Ryle's own criteria, in my idiolect at least, 'exist' has not been shown to be ambiguous. I do not even hear any linguistic absurdity in:

(6) *Minds and bodies exist in space and time.*

This Cartesian statement is surely false, but even if Cartesian minds were essentially non-spatial-temporal, the statement that they are spatio-temporal is no more linguistically absurd than the statement that seven is an even integer. The latter is a priori false (if anything is), but, for all that, it is not linguistically absurd. So far, then, Morton White's, and J. S. Mill's, univocalism about 'exist' is sustained.

[8] As Lyons (1977: 407) points out, and I agree, this inference is not justified. Zeugma does not entail ambiguity. See ch. 2, sect. 1.

[9] Assumption (ii) is an extension of the version he gave in Ryle (1945: 2), "Knowing How and Knowing That": if Type (*e*) ≠ Type (*e'*), then *A(e)* is linguistically acceptable only if *A(e')* is unacceptable. In Ryle (1938: 294–5), "Categories", he had asserted: if Type(*e*) = Type(*e'*), then *A(e)* is acceptable only if *A(e')* is acceptable, a variant of (i).

Certain features of these familiar, historical arguments seem to me worth emphasizing. Ryle, like Russell, draws inferences from two kinds of semantic data, logical inconsistency and linguistic anomaly. But Ryle, unlike Russell, distinguishes between the two kinds of data. Furthermore, Ryle, like Aristotle, explicitly offers linguistic argumentation. From the alleged anomaly of conjoined noun phrases Ryle concludes that an expression in whose scope they occur must be ambiguous. This argumentation is in the service of Ryle's notable epistemological and metaphysical theses. It seems to me worth examining the cogency of this argumentation. Astonishingly little serious examination has ever been done, and in a Chomskyan intellectual world it seems to me that we should try to assess what Russell, Ryle, and White, not to mention Aristotle and John Stuart Mill, have accomplished. The success of their epistemology and metaphysics depends upon the correctness of their linguistics.

Bibliography

AKMAJIAN, A., and HENY, F. (1975). *An Introduction to the Principles of Transformational Syntax*. Cambridge: MIT Press.

ALLWOOD, J. (1972). "Negation and the Strength of Presupposition". *Logical Grammar Reports No. 2*. Göteborg: Dept. of Linguistics.

—— (1977). "Negation and the Strength of Presupposition", rev. edn. in O. Dahl (ed.), *Logic, Pragmatics, and Grammar*. Göteborg: Dept. of Linguistics. Pp. 11–52.

ANSCOMBE, G. E. M. (1981). *Metaphysics and the Philosophy of Mind*. Minneapolis: Univ. of Minn. Press.

ATLAS, J. D. (1972). "Some Remarks on G. Lakoff's 'Performative Antinomies'", Claremont: Pomona College. TS.

—— (1974). "Presupposition, Ambiguity, and Generality: A Coda to the Russell–Strawson Debate on Referring", Claremont: Pomona College.

—— (1975*a*). "Frege's Polymorphous Concept of Presupposition and its Role in a Theory of Meaning", *Semantikos*, 1: 29–44.

—— (1975*b*). "Presupposition: A Semantico-Pragmatic Account", *Pragmatics Microfiche*, I. 4: D13–G9.

—— (1977*a*). "Presupposition Revisited", *Pragmatics Microfiche*, II. 5: D5–D11.

—— (1977*b*). "Negation, Ambiguity, and Presupposition", *Linguistics and Philosophy*, 1: 321–36.

—— (1978). "On Presupposing", *Mind*, 87: 396–411.

—— (1979). "How Linguistics Matters to Philosophy: Presupposition, Truth, and Meaning", in D. Dinneen and C.-K. Oh (eds.), *Syntax and Semantics* 11: *Presupposition*. New York: Academic Press. Pp. 265–81.

—— (1980*a*). "A Note on a Confusion of Pragmatic and Semantic Aspects of Negation", *Linguistics and Philosophy*, 3: 411–14.

—— (1980*b*). "Reference, Meaning, and Translation", *Philosophical Books*, 21: 129–40.

—— (1981). "Is *not* logical?", in *Proceedings: The Eleventh International Symposium on Multiple-Valued Logic*. New York: The Institute of Electrical and Electronics Engineers Computer Society Press. Pp. 124–8.

—— (1982). "Non-Existence and Quantifying in". Univ. of California, Los Angeles, Colloquium Lecture. TS.

—— (1984*a*). "Comparative Adjectives and Adverbials of Degree:

An Introduction to Radically Radical Pragmatics", *Linguistics and Philosophy*, 7: 347–77.

ATLAS, J. D. (1984*b*). "Grammatical Non-Specification: The Mistaken Disjunction Theory", *Linguistics and Philosophy*, 7: 433–43.

—— (1988*a*). Review of W. G. Lycan, *Logical Form in Natural Language*, (1984), *Language*, 64: 158–67.

—— (1988*b*). "What Are Negative Existence Statements About?", *Linguistics and Philosophy*, 11: 371–93.

—— and LEVINSON, S. C. (1981). "*It*-Clefts, Informativeness, and Logical Form: Radical Pragmatics (Revised Standard Version)", in P. Cole (ed.), *Radical Pragmatics*. New York: Academic Press. Pp. 1–61.

ATTNEAVE, F. (1971). "Multistability in Perception", *Scientific American*, Dec. 1971: 91–9.

AYER, A. J. (1952). "Negation", *Journal of Philosophy*, 49: 797–815. Repr. in Ayer (1954): 36–65.

—— (1954). *Philosophical Essays*. London: Macmillan.

BACH, K. (1982). "Semantic Nonspecificity and Mixed Quantifiers", *Linguistics and Philosophy*, 4: 593–605.

—— and Harnish, M. (1979). *Linguistic Communication and Speech Acts*. Cambridge: MIT Press.

BARWISE, J., and PERRY, J. (1983). *Situations and Attitudes*. Cambridge: MIT Press.

BENNET, J. (1971). *Locke, Berkeley, Hume*. Oxford: Clarendon Press.

BERGMANN, M. (1977). "Logic and Sortal Incorrectness", *Review of Metaphysics*, 31: 61–79.

—— (1981). "Presupposition and Two-Dimensional Logic", *Journal of Philosophical Logic*, 10: 27–53.

BLACK, M.(1946). "Russell's Philosophy of Language", in P. A. Schilpp (ed.), *The Philosophy of Bertrand Russell*. Evanston, Ill.: Lib. of Living Philosophers. Pp. 229–55.

—— (1972). "How Do Pictures Represent?", in E. H. Gombrich *et al*, *Art, Perception, and Reality*. Baltimore: Johns Hopkins Univ. Press. Pp. 95–130.

BLACKBURN, W. K. (1983). "Ambiguity and Non-specificity: A Reply to Jay David Atlas", *Linguistics and Philosophy*, 4: 493–605.

BLOOMfIELD, L (1933). *Language*. New York: Holt.

BOCHVAR, D. A. (1939). "Ob odnom tréhnačom isčislenii i égo primeńénii K analizu paradoksov klassičéskogo rasširénnogo funkcional'nogo isčisléniá", *Matematičéskij sbornik*, 4: 287–308. (Reviewed in Church 1939, 1940).

BODEN, M. A. (1988). *Computer Models of Mind*. Cambridge: Camb. Univ. Press.

BOËR, S (1979). "Meaning and Contrastive Stress", *Philosophical Review*, 77: 263–98.

—— and LYCAN, W. G. (1976). *The Myth of Semantic Presupposition*. Bloomington: Indiana Univ. Linguistics Club.

BRADLEY, F. H. (1883). *Principles of Logic*. Oxford: Clarendon Press.

BRONOWSKI, J. (1978). *The Origins of Knowledge and Imagination*. New Haven: Yale Univ. Press.

BRUGMAN, C. (1981). *The Story of 'Over'*. Bloomington: Indiana Univ. Linguistics Club.

BURGE, T. (1973). "Reference and Proper Names", *Journal of Philosophy*, 70: 425–39.

CARSTON, R. (1985*a*). "A Reanalysis of Some 'Quantity Implicatures' ". Univ. of London. TS.

—— (1985*b*). "Saying and Implicating". Univ. of London. TS.

CHOMSKY, N. (1957). *Syntactic Structures*. The Hague: Mouton.

—— (1959). Review of B. F. Skinner, "Verbal Behavior", *Language*, 35: 26–8.

—— (1972). *Language and Mind*. New York: Harcourt Brace Jovanovich.

—— (1986). *Knowledge of Language*. New York: Praeger.

CHURCH, A (1939/40). Review of: D. A. Bochvar, "On a Three-valued Logical Calculus and its Application to the Analysis of Contradictions", *Journal of Symbolic Logic*, 4: 98–9; 5: 119.

COHEN, L. J. (1962). *Diversity of Meaning*. London: Methuen.

—— (1971). "The Logical Particles of Natural Language", in Y. Bar-Hillel (ed.), *Pragmatics of Natural Language*. Dordrecht: Reidel.

—— (1980). "The Individuation of Proper Names", in Z. van Straaten (ed.), *Philosophical Subjects*. Oxford: Clarendon Press. Pp. 287–94.

—— (1985). "A Problem about Ambiguity in Truth-Theoretical Semantics", *Analysis*, 45: 129–34.

—— (1986). "How is Conceptual Innovation Possible?", *Erkenntnis*, 25: 221–38.

CRUSE, D. A. (1986). *Lexical Semantics*. Cambridge: Camb. Univ. Press.

DAVIDSON, D. (1967*a*). "The Logical Form of Action Sentences", in N. Rescher (ed.), *The Logic of Decision and Action*. Pittsburgh: Univ. of Pittsburgh Press. Pp. 81–95.

—— (1967*b*). "Truth and Meaning", *Synthese*, 17: 304–23.

—— (1970). "Semantics for Natural Languages", in B. Visentini *et al.* (eds.), *Linguaggi nella società e nella tecnica*. Milano: Edizioni di Comunita. Pp. 177–88. Repr. in Davidson (1984), pp. 55–64.

—— (1984). *Inquiries into Truth and Interpretation*. Oxford: Clarendon Press.

—— and HARMAN, G. (eds.) (1972). *The Semantics of Natural Language*. Dordrecht: Reidel.

—— —— (eds.) (1975). *The Logic of Grammar*. Encino: Dickinson.

DEUTSCH, D. (1987). "The Tritone Paradox: Effects of Spectral Variables", *Perception and Psychophysics*, 41: 563–75.

DEUTSCH, D. and FEROE, J. (1981). "The Internal Representation of Pitch Sequences in Tonal Music", *Psychological Review*, 88: 503–22.

DILLON, G. L. (1977). *Introduction to Contemporary Linguistic Semantics*. Englewood Cliffs, NJ: Prentice-Hall.

DINSMORE, J. (1981). "Towards a Unified Theory of Presupposition", *Journal of Pragmatics*, 5: 335–63.

DONNELLAN, K. (1962). "Necessity and Criteria", *Journal of Philosophy*, 59: 647–58.

—— (1975). "Speaking of Nothing", in D. Hockney *et al.* (eds.), *Contemporary Research in Philosophical Logic and Linguistic Semantics*. Dordrecht: Reidel. Pp. 93–118.

—— (1978). "Speaker Reference, Descriptions, and Anaphora", in P. Cole (ed.), *Syntax and Semantics 9: Pragmatics*. New York: Academic Press. Pp. 47–68. Repr. in French *et al.* (1979), pp. 28–44.

—— (1981). "Intuitions and Presuppositions", in P. Cole (ed.), *Radical Pragmatics*. New York: Academic Press. Pp. 129–42.

—— (1983). "Kripke and Putnam on Natural Kind Terms", in C. Ginet and S. Shoemaker (eds.), *Knowledge and Mind*. Oxford: Clarendon Press. Pp. 84–104.

DOWTY, D., WALL, R., and PETERS, S. (1981). *Introduction to Montague Semantics*. Dordrecht: Reidel.

DRESHER, E. (1977). "Logical Representations and Linguistic Theory", *Linguistic Inquiry*, 8: 351–78.

DUMMETT, M. (1959). "Truth", *Proceedings of the Aristotelian Society*, 59: 141–62. Repr. in Dummett (1978), pp. 1–24.

—— (1969). "The Reality of the Past", *Proceedings of the Aristotelian Society*, 69: 239–58. Repr. in Dummett (1978), pp. 358–74.

—— (1973). "The Justification of Deduction", *Proceedings of the British Academy*, 59: 3–34. Repr. in Dummett (1978), pp. 290–318.

—— (1978). *Truth and other Enigmas*. London: Duckworth.

EJERHED, E. (1981). "Tense as a Source of Intensional Ambiguity", in F. Heny (ed.), *Ambiguities in Intensional Contexts*. Dordrecht: Reidel. Pp. 231–52.

EMPSON, W. (1930). *Seven Types of Ambiguity*. London: Chatto and Windus.

FODOR, J. A. (1983*a*). "Imagery and the Language of Thought", in J. Miller (ed.), *States of Mind*. New York: Pantheon. Pp. 84–98.

—— (1983*b*). *The Modularity of Mind*. Cambridge: MIT Press.

FODOR, J. D. (1977). *Semantics: Theories of Meaning in Generative Grammar*. Hassocks: Harvester Press.

FOSTER, L. (1971). "Hempel, Scheffler, and the Ravens", *Journal of Philosophy*, 68: 107–14.

FRENCH, P. A., VEHLING, jun., T. E., and WETTSTEIN, H. K. (eds.) (1979).

Contemporary Perspectives in the Philosophy of Language. Minneapolis: Univ. of Minn. Press.

FREGE, G. (1884). *The Foundations of Arithmetic*, trans. J. L. Austin, 1950. Oxford: Blackwell.

GAZDAR, G. (1979). *Pragmatics: Implicature, Presupposition, and Logical Form*. New York: Academic Press.

—— (1980). "Pragmatics and Logical Form", *Journal of Pragmatics*, 4: 1–13.

GAZZANIGA, M. (1985). *The Social Brain*. New York: Basic Books.

GOGUEN, J. A. (1969). "The Logic of Inexact Concepts", *Synthese*, 19: 325–73.

GOMBRICH, E. H. (1969). *Art and Illusion*. Princeton: Princeton Univ. Press.

—— (1972). "The Mask and the Face", in E. Gombrich *et al.*, *Art, Perception, and Reality*. Baltimore: Johns Hopkins Univ. Press. Pp. 1–46.

GOODMAN, N. (1949). "On Likeness of Meaning", *Analysis*, 10: 67–74. Repr. in Goodman (1972), pp. 221–30.

—— (1972). *Problems and Projects*. Indianapolis: Hackett.

—— (1978). *Ways of Worldmaking*. Indianapolis: Hackett.

—— (1983). *Fact, Fiction, and Forecast*. Cambridge: Harvard Univ. Press.

—— and ELGIN, C. Z. (1988). *Reconceptions in Philosophy and in other Arts and Sciences*. Indianapolis: Hackett.

GRANDY, R. (1974). "Some Remarks About Logical Form", *Noûs*, 8: 157–64.

—— (1987). "In Defence of Semantic Fields", in E. LePore (ed.), *New Directions in Semantics*. New York: Academic Press. Pp. 259–80.

—— (1988). "Theories of Truth and Ambiguities". Houston: Rice Univ. TS.

GREEN, G. (1969). "On the Notion of 'Related Lexical Entry'", *Papers from the Fifth Regional Meeting of the Chicago Linguistic Society*. Chicago: Chi. Linguistic Society. Pp. 76–88.

GREGORY, R. L. (1970). *The Intelligent Eye*. London: Weidenfeld and Nicolson.

—— (1973). "The Confounded Eye", in R. L. Gregory and E. H. Gombrich, *Illusion in Nature and Art*. London: Duckworth. Pp. 49–95.

—— (1986). *Odd Perceptions*. London: Methuen.

GRICE, H. PAUL (1961). "The Causal Theory of Perception", *Proceedings of the Aristotelian Society*, 25: 121–52.

—— (1975). "Logic and Conversation", in P. Cole and J. L. Morgan (eds.), *Syntax and Semantics 3: Speech Acts*. New York: Academic Press. Pp. 41–58.

GRICE, H. PAUL (1978). "Further Notes on Logic and Conversation", in P. Cole (ed.), *Syntax and Semantics 9: Pragmatics*. New York: Academic Press. Pp. 113–27.

—— (1981). "Presupposition and Conversational Implicature", in P. Cole (ed.), *Radical Pragmatics*. New York: Academic Press. Pp. 183–97.

GRINDER, J., and POSTAL, P. M. (1971). "Missing Antecedents", *Linguistic Inquiry*, 2: 269–312.

GUNDEL, J. (1977). *Role of Topic and Comment in Linguistic Theory*. Bloomington: Indiana Univ. Linguistics Club.

HARMAN, G. (1970). "Deep Structure as Logical Form", *Synthese*, 21: 275–97.

—— (1973). "Review of *Language and Mind*", *Language*, 49: 453–64.

—— (1974). "First General Discussion Session", *Synthese*, 27: 471–508.

HERZBERGER, H. G. (1973). "Dimensions of Truth", *Journal of Philosophical Logic*, 2: 535–56.

—— (1975). "Supervaluations in Two-Dimensions". *Proceedings: The Fifth International Symposium on Multiple-Valued Logic*. New York: Institute of Electrical and Electronics Engineers Computer Society Press. Pp. 124–8.

HINTIKKA, J (1986). "Logic of Conversation as a Logic of Dialogue", in R. E. Grandy and R. Warner (eds.), *Philosophical Grounds of Rationality*. Oxford: Clarendon Press. Pp. 259–76.

HOCHBERG, J. (1972). "The Representation of Things and People", in E. H. Gombrich *et al.*, *Art, Perception, and Reality*. Baltimore: Johns Hopkins Univ. Press. Pp. 47–94.

HOLLINGS, L. (1980). "Presuppositions and Theories of Meaning", *Mind*, 89: 274–81.

HORN, L. (1978*a*). "Some Aspects of Negation", in J. Greenberg *et al.* (eds.), *Universals of Human Language 4: Syntax*. Stanford: Stanford Univ. Press. Pp. 127–210.

—— (1978*b*). "Remarks on Neg-Raising", in P. Cole (ed.), *Syntax and Semantics 9: Pragmatics*. New York: Academic Press. Pp.129–220.

—— (1981). "Exhaustiveness and the Semantics of Clefts", *Proceedings of the New England Linguistic Society*, 11: 125–42.

—— (1984*a*). "Ambiguity, Negation, and the London School of Parsimony", *Proceedings of the New England Linguistic Society*, 14: 108–31.

—— (1984*b*). "Toward a New Taxonomy for Pragmatic Inference: Q-Based and R-Based Implicature", in D. Schiffrin (ed.), *Meaning, Form, and Use in Context: Linguistic Applications (Georgetown University Round Table '84)*. Washington: Georgetown Univ. Press. Pp. 11–42.

—— (1985). "Metalinguistic Negation and Pragmatic Ambiguity", *Language*, 61: 121–74.

—— (1988). "Pragmatic Theory", in F. J. Newmeyer (ed.), *Linguistics: The Cambridge Survey; vol. i. Linguistic Theory—Foundations*. Cambridge: Camb. Univ. Press. Pp. 113–45.
—— (1989). *A Natural History of Negation*. Chicago: Univ. of Chi. Press.
HYMAN, L. M. (1975). *Phonology: Theory and Analysis*. New York: Holt, Rinehart, and Winston.
JACKENDOFF, R. (1983). *Semantics and Cognition*. Cambridge: MIT Press.
JOHNSON, M. (1987). *The Body in the Mind*. Chicago: Univ. of Chi. Press.
KAPLAN, D. (1979). "On the Logic of Demonstratives", in P. French *et al.* (eds.), *Contemporary Perspectives in the Philosophy of Language*. Minneapolis: Univ. of Minn. Press. Pp. 401–12.
KARTUNNEN, L. (1971*a*). "Implicative Verbs", *Language*, 47: 340–58.
—— (1971*b*). "Some Observations on Factivity", *Papers in Linguistic*, 4: 55–70.
—— (1973*a*). "Presuppositions of Compound Sentences", *Linguistic Inquiry*, 4: 169–93.
—— (1973*b*). "Remarks on Presuppositions", Austin: Univ. of Texas. TS.
KATZ, J. J. (1972). *Semantic Theory*. New York: Harper and Row.
—— (1980). "Chomsky on Meaning", *Language*, 56: 1–41.
—— (1981). *Language and Other Abstract Objects*. Oxford: Blackwell.
—— (1986). *Cogitations*. Oxford: Clarendon Press.
—— (1987). "Common Sense in Semantics", in E. LePore (ed.), *New Directions in Semantics*. New York: Academic Press. Pp. 157–233.
—— and Langendoen, D. T. (1976). "Pragmatics and Presupposition", *Language*, 52: 1–17.
KEMPSON, R. (1975). *Presupposition and the Delimitation of Semantics*. Cambridge: Camb. Univ. Press.
—— (1977). *Semantic Theory*. Cambridge: Camb. Univ. Press.
—— (1979). "Presupposition, Opacity, and Ambiguity", in C-K Oh and D. Dinneen (eds.), *Syntax and Semantics 11: Presupposition*. New York: Academic Press. Pp. 283–98.
—— (1986). "Ambiguity and the Semantics-Pragmatics Distinction", in C. Travis (ed.), *Meaning and Interpretation*. Oxford: Blackwell. Pp. 77–103.
—— (1988). "Grammar and Conversational Principles", in F. J. Newmeyer (ed.), *Linguistics: The Cambridge Survey*; *vol. ii. Linguistic Theory—Extensions and Implications*. Cambridge: Camb. Univ. Press. Pp. 139–63.
—— and CORMACK, A. (1981). "Ambiguity and Quantification", *Linguistics and Philosophy*, 4: 259–309.
—— (1982). "Quantification and Pragmatics", *Linguistics and Philosophy*, 4: 607–18.
KIPARSKY, P., and KIPARSKY, C. (1970). "Fact", in M. Bierwisch and

K. Heidolph (eds.), *Progress in Linguistics*. The Hague: Mouton. Pp. 143–73. Repr. in D. Steinberg and L. Jakobovits (eds.) (1971) *Semantics*. Cambridge: Camb. Univ. Press. Pp. 345–69.

Kleene, S. C. (1938). "On a Notation for Ordinal Numbers", *Journal of Symbolic Logic*, 3: 150–5.

—— (1971). *Introduction to Metamathematics*. Groningen: Wolters-Noordhoof. Pp. 332–40.

Klima, E. (1964). "Negation in English", in J. A. Fodor and J. Katz (eds.), *The Structure of Language*. Englewood Cliffs: Prentice-Hall. Pp. 246–323.

Kneale, W., and Kneale, M. (1962). *The Development of Logic*. Oxford: Clarendon Press.

Kooij, J. (1971). *Ambiguity in Natural Language*. Amsterdam: North-Holland.

Kripke, S. (1977). "Speaker's Reference and Semantic Reference", in P. A. French, T. E. Uehling, jun., and H. K. Wettstein (eds.), *Midwest Studies in Philosophy*. Minneapolis: Univ. of Minn. Press. Pp. 255–76. Repr. in French *et al.* (1979), pp. 6–27.

—— (1982). *Wittgenstein on Rules and Private Language*, Cambridge: Harvard Univ. Press.

Lakoff, G. (1966). "Deep and Surface Grammar." Bloomington: Indiana Univ. Linguistics Club.

—— (1970). "A Note on Ambiguity and Vagueness", *Linguistic Inquiry*, 1: 357–9.

—— (1971). "On Generative Semantics", in D. Steinberg and L. Jakobovits (eds.), *Semantics*. Cambridge: Camb. Univ. Press. Pp. 232–96.

—— (1972). "Performative Antinomies", *Foundations of Language*, 8: 569–72.

—— (1975). "Pragmatics in Natural Logic", in E. L. Keenan (ed.), *Formal Semantics of Natural Language*. Cambridge: Camb. Univ. Press. Pp. 253–86.

—— (1977). "Linguistic Gestalts", *Papers from the Thirteenth Regional Meeting, Chicago Linguistic Society*. Chicago: Chi. Linguistic Society. Pp. 236–87.

—— (1982). *Categories and Cognitive Models*. Berkeley: Univ. of Calif. Institute of Human Learning.

—— (1986). "Two Metametaphorical Issues", *Berkeley Cognitive Science Report No. 38* (May 1986). Berkeley: Univ. of Calif. Institute of Cognitive Studies.

—— (1987). *Women, Fire, and Dangerous Things: What Categories Reveal About the Mind*. Chicago: Univ. of Chi. Press.

Larson, R. (1988). "Scope and Comparatives", *Linguistics and Philosophy*, 11: 1–26.

LEECH, G. N. (1983). *Principles of Pragmatics*. London: Longman.

LEHRER, A. (1974). *Semantic Fields and Lexical Structure*. Amsterdam: North-Holland.

LEPORE, E. (ed.) (1987). *New Directions in Semantics*. New York: Academic Press.

LEVINSON, S. C. (1983). *Pragmatics*. Cambridge: Camb. Univ. Press.

—— (1985). "Minimization and Conversational Inference", in M. Papi and J. Verscheuren (eds.), *The Pragmatic Perspective: Proceedings of the International Pragmatics Conference, Viareggio, 1985*. Amsterdam: Benjamin. Pp. 61–129.

—— (1987). "Pragmatics and the Grammar of Anaphora: a Partial Pragmatic Reduction of Binding and Control Phenomena", *Journal of Linguistics*, 23: 379–434.

—— (1988). "Generalized Conversational Implicatures and the Semantics/Pragmatics Interface". Univ. of Cambridge. TS.

—— (forthcoming). *Implicature*. Cambridge: Cambridge Univ. Press.

LEWIS, D. K. (1972). "General Semantics", in G. Harman and D. Davidson (eds.), *Semantics of Natural Language*. Dordrecht: Reidel. Pp. 169–218.

—— (1982). "Logic for Equivocators", *Noûs*, 16: 431–41.

LIGHTFOOT, D. (1982). *The Language Lottery: Toward a Biology of Grammars*. Cambridge: MIT Press.

LIZOTTE, R. J. (1983). "Universals Concerning Existence, Possession, and Location Sentences", Ph.D. Diss., Brown University.

LUKASIEWICZ, J. (1967). "Philosophical Remarks on Many-Valued Systems of Propositional Logic", in S. McCall (ed.), *Polish Logic: 1920–1939*. Oxford: Clarendon Press. Pp. 40–65.

LUNTLEY, M. (1988). *Language, Logic, and Experience*. La Salle: Open Court.

LYCAN, W. G. (1984). *Logical Form in Natural Language*. Cambridge: MIT Press.

LYONS, J. (1977). *Semantics*. Cambridge: Camb. Univ. Press.

MACHINA, K. (1972). "Vague Predicates", *American Philosophical Quarterly*, 9: 225–33.

MACKIE, J. L. (1976). *Problems from Locke*. Oxford: Clarendon Press.

MANSER, A. (1983). *Bradley's Logic*. Oxford: Blackwell.

MARGALIT, A. (1983). Review of: I. Scheffler, *Beyond the Letter: A Philosophical Inquiry into Ambiguity, Vagueness, and Metaphor in Language* (Boston, 1979), *Journal of Philosophy*, 80: 129–38.

MARTIN, J. N. (1975). "Karttunen on Possibility", *Linguistic Inquiry*, 6: 339–41.

—— (1979). "Some Misconceptions in the Critique of Semantic Presupposition", *Theoretical Linguistics*, 6: 235–82.

—— (1982*a*). "Negation, Ambiguity, and the Identity Test", *Journal of Semantics*, 1: 251–74.

MARTIN, J. N. (1982*b*). "The Role of Negation in Presupposition Theory", in *Proceedings: The Twelfth International Symposium on Multiple-Valued Logic*. New York: The Institute of Electrical and Electronics Engineers Computer Society Press. Pp. 323–8.

—— (1987). *Elements of Formal Semantics*. New York: Academic Press.

MATES, B. (1952). "Synonymity", in L. Linsky (ed.), *Semantics and the Philosophy of Language*. Urbana: Univ. of Ill. Press. Pp. 111–36.

MCCAWLEY, J. D. (1968). "The Role of Semantics in a Grammar", in E. Bach and R. T. Harms (eds.), *Universals in Linguistic Theory*. New York: Holt, Rinehart, and Winston. Pp. 125–70.

—— (1981). *Everything that Linguists have always wanted to know about Logic: but were ashamed to ask*. Chicago: Univ. of Chi. Press.

MCDOWELL, J. (1976). "Truth Conditions, Bivalence, and Verificationism", in G. Evans and J. McDowell (eds.), *Truth and Meaning: Essays in Semantics*. Oxford: Clarendon Press. Pp. 42–66.

MILLER, G., and JOHNSON-LAIRD, P. N. (1976). *Language and Perception*. Cambridge: Harvard Univ. Press.

MONTAGUE, R. (1970). "Universal Grammar", *Theoria*, 36: 373–98. Repr. in R. Montague (1974), pp. 222–46.

—— (1974). *Formal Philosophy*, ed. R. Thomason. New Haven: Yale Univ. Press.

NEWMEYER, F. J. (1986). *Linguistic Theory in America*, 2nd edn. New York: Academic Press.

—— (1988). *Linguistics: The Cambridge Survey*. Cambridge: Camb. Univ. Press.

NIDDITCH, P. H. (1980). *Draft A of Locke's Essay concerning Human Understanding: The earliest extant Autograph Version, transcribed with Critical Apparatus*. Sheffield: Univ. of Sheffield.

ORNSTEIN, R. (1986). *Multimind*. Boston: Houghton Mifflin.

PARSONS, K. P. (1973). "Ambiguity and the Truth Definition", *Noûs*, 7: 379–94.

PARSONS, T. (1980). *Nonexistent Objects*. New Haven: Yale Univ. Press.

PASSMORE, J. (1970). *Philosophical Reasoning*. New York: Basic Books.

PARTEE, B. H. (1979). "Montague Grammar, Mental Representations, and Reality", in P. A. French *et al.* (eds.), *Contemporary Perspectives in the Philosophy of Language*. Minneapolis: Univ. of Minn. Press. Pp. 195–208. Repr. in S. Kanger and S. Ohmann (eds.) (1981), *Philosophy and Grammar: Papers on the Occasion of the Quincentennial of Uppsala University*. Dordrecht: Reidel. Pp. 59–78.

PEARS, D. (1967). "Is Existence a Predicate?", in P. F. Strawson (ed.), *Philosophical Logic*. Oxford: Clarendon Press. Pp. 97–106.

PULMAN, S. G. (1983). *Word Meaning and Belief*. London: Croom Helm.

PUTNAM, H. (1958). "Formalization of the Concept 'About' ", *Philosophy of Science*, 25: 125–30.

—— (1973). "Meaning and Reference", *Journal of Philosophy*, 70: 699–711.

—— (1975). "The Meaning of 'Meaning'", in K. Gunderson (ed.), *Language, Mind, and Knowledge*. Minneapolis: Univ. of Minn. Press. Pp. 131–93.

QUINE, W. V. O. (1960). *Word and Object*, Cambridge: MIT Press.

—— (1969). "Natural Kinds", in his *Ontological Relativity and Other Essays*. New York: Columbia Univ. Press.

—— (1974). "First General Discussion Session", *Synthese*, 27: 471–508.

—— (1976). *The Ways of Paradox and Other Essays*. Cambridge: Harvard Univ. Press.

—— (1980). *From a Logical Point of View*. Cambridge: Harvard Univ. Press.

RADFORD, A. (1981). *Transformational Syntax*. Cambridge: Camb. Univ. Press.

REEVES, A. (1975). "Ambiguity and Indifference", *Australasian Journal of Philosophy*, 53: 220–37.

RESCHER, N. (1969). "*Many-Valued Logic*. New York: McGraw-Hill.

RICHMAN, R. (1959). "Ambiguity and Intuition", *Mind*, 68: 87–92.

RIEMSDIJK, H. van, and WILLIAMS, E. (1986). *Introduction to the Theory of Grammar*. Cambridge: MIT Press.

ROBERTS, L. D. (1984). "Ambiguity vs. Generality: Removal of a Logical Confusion", *Canadian Journal of Philosophy*, 14: 295–313.

—— (1987). "Intuitions and Ambiguity Tests", *Canadian Journal of Philosophy*, 17: 189–98.

ROSS, J. F. (1981). *Portraying Analogy*. Cambridge: Camb. Univ. Press.

ROSS, J. R. (1967). "Constraints on Variables in Syntax". Ph.D. Diss., MIT.

RUHL, C. S. (1975). "Polysemy or Monosemy: Discrete Meanings or Continuum", in R. W. Fasold and R. W. Shuy (eds.), *Analyzing Variation in Language: Papers from the Second Colloquium on New Ways of Analyzing Variation*. Washington: Georgetown Univ. Press. Pp. 184–202.

RUSSELL, B. (1903). *Principles of Mathematics*. Cambridge: Camb. Univ. Press.

—— (1905). "On Denoting", *Mind*, 14: 479–93. Repr. in Russell (1956), pp. 41–56.

—— (1908). "Mathematical Logic as Based on the Theory of Types", *American Journal of Mathematics*, 30: 222–62. Repr. in Russell (1956), p. 59–102.

—— (1918). "The Philosophy of Logical Atomism", *Monist*, 28: 495–527. Repr. in Russell (1956), pp. 178–203.

—— (1919*a*). "The Philosophy of Logical Atomism", *Monist*, 29: 32–63, 190–222, 345–80. Repr. in Russell (1956), pp. 203–81.

RUSSELL, B. (1919*b*). *Introduction to Mathematical Philosophy*. London: George Allen and Unwin.

—— (1924). "Logical Atomism", in J. H. Muirhead (ed.), *Contemporary British Philosophy*. 1st ser. London: George Allen and Unwin. Pp. 356–83. Repr. in Russell (1956), pp. 323–43.

—— (1945). *A History of Western Philosophy*. New York: Simon and Schuster.

—— (1946). "Replies to Criticisms", in P. A. Schilpp (ed.), *The Philosophy of Bertrand Russell*. Evanston, Ill.: Lib. of Living Philosophers.

—— (1956). *Logic and Knowledge: Essays 1901–1950*, ed. R. Marsh. London: George Allen and Unwin.

—— (1973). *Essays in Analysis*, ed. D. Lackey. New York: George Braziller.

—— and WHITEHEAD, A. N. (1910). *Principia Mathematica*, vol. I. Cambridge: Camb. Univ. Press.

—— —— (1927). *Principia Mathematica*, vol. I. 2nd edn. Cambridge: Camb. Univ. Press.

RYLE, G. (1938). "Categories", *Proceedings of the Aristotelian Society*, 38: 189–206.

—— (1945). "Knowing How and Knowing That", *Proceedings of the Aristotelian Society*, 46: 1–16.

—— (1949). *Concept of Mind*. London: Hutchinson.

SADOCK, J. (1975). "Larry Scores a Point", *Pragmatics Microfiche*, I. 4: G10–G14.

—— (1978). "On Testing for Conversational Implicature", in P. Cole (ed.), *Syntax and Semantics 9: Pragmatics*. New York: Academic Press. Pp. 281–97.

—— and ZWICKY, A. (1987). "A Non-test for Ambiguity", *Canadian Journal of Philosophy*, 17: 185–7.

SCHEFFLER, I. (1979). *Beyond the Letter: A Philosophical Inquiry into Ambiguity, Vagueness, and Metaphor in Language*. Boston: Routledge and Kegan Paul.

SCOTT, DANA S. (1973). "Background to Formalization", in H. Leblanc (ed.), *Truth, Syntax, and Modality*. Amsterdam: North-Holland. Pp. 244–73.

—— (1976). "Does Many Valued Logic Have Any Use?", in S. Körner (ed.), *Philosophy of Logic*. Berkeley: Univ. of Calif. Press. Pp. 64–74.

SELLARS, W. (1963). *Science, Perception, and Reality*. New York: Humanities.

SEUREN, P. (1985). *Discourse Semantics*. Oxford: Blackwell.

SMILEY, T. J. (1976). "Comment", in S. Körner (ed.), *Philosophy of Logic*. Berkeley: Univ. of Calif. Press. Pp. 74–88.

SPERBER, D., and WILSON, D. (1986). *Relevance*. Cambridge: Harvard Univ. Press.

Stalnaker, R. (1972). "Pragmatics", in D. Davidson and G. Harman (eds.), *Semantics of Natural Language*. Dordrecht: Reidel. Pp. 380–97.

Staniland, H. (1972). *Universals*. London: Macmillan.

Strawson, P. F. (1950). "On Referring", *Mind*, 59: 320–44. Repr. in Strawson (1971), pp. 1–27.

—— (1952). *Introduction to Logical Theory*. London: Methuen.

—— (1954). "A Reply to Mr Sellars", *Philosophical Review*, 63: 216–31.

—— (1964). "Identifying Reference and Truth-Values", *Theoria*, 30: 96–118. Repr. in Strawson (1971), pp. 75–95.

—— (1971). *Logico-Linguistic Papers*. London: Methuen.

—— (1980). "Reply to Cohen, Quine, and Geach", in Z. van Straaten (ed.), *Philosophical Subjects*. Oxford: Clarendon Press. Pp. 287–94.

Tarski, A. (1936). "Der Wahrheitsbegriff in den formalisierten Sprachen", *Studia Philosophica*, 1: 261–405. Repr. in Tarski (1956), pp. 152–278.

—— (1956). *Logic, Semantics, Metamathematics*. Oxford: Clarendon Press.

Tennant, N. (1981). "Is This a Proof I See Before Me?", *Analysis*, 41: 115–19.

Thomason, R. (1973). "Semantics, Pragmatics, Conversation, and Presupposition", Pittsburgh: Univ. of Pittsburgh. TS.

van der Sandt, R. A. (1988). *Context and Presupposition*. London: Croom Helm.

van Fraassen, B. (1966). "Singular Terms, Truth-value Gaps, and Free Logic", *Journal of Philosophy*, 63: 481–95.

—— (1969). "Presuppositions, Supervaluations, and Free Logic", in K. Lambert (ed.), *The Logical Way of Doing Things*. New Haven: Yale Univ. Press. Pp. 67–91.

—— (1971). *Formal Logic and Semantics*. New York: Macmillan.

Waismann, F. (1945–6). "Are There Alternative Logics?", *Proceedings of the Aristotelian Society, Supplementary Volume 19*: 119–50.

Watson, J. B. (1924). *Behaviorism*. New York: W. W. Norton.

Wertheimer, R. (1972). *The Significance of Sense: Meaning, Modality, and Morality*. Ithaca: Cornell Univ. Press.

Weydt, H. (1972). "Le Concept d'ambiguité en grammaire transformationelle-générative et en linguistique fonctionelle", *La Linguistique*, 8: 41–72.

—— (1973). "On G. Lakoff, 'Instrumental Adverbs and the Concept of Deep Structure'", *Foundations of Language*, 10: 569–78.

White, M. G. (1956/63). *Toward Reunion in Philosophy*. Cambridge: Harvard Univ. Press.

Wilson, D. (1975). *Presuppositions and Non-truth-conditional Semantics*. New York: Academic Press.

Wittgenstein, L. (1958). *Philosophical Investigations*. Oxford: Blackwell.

—— (1969). *On Certainty*. Oxford: Blackwell.

WITTGENSTEIN, L. (1970). *Zettel*. Berkeley: Univ. of California Press.

—— (1978). *Remarks on the Foundations of Mathematics*. Oxford: Blackwell.

WRIGHT, C. (1980). *Wittgenstein on the Foundations of Mathematics*. London: Duckworth.

—— (1987). *Realism, Meaning, and Truth*. Oxford: Blackwell.

ZIFF, P. (1972). *Understanding Understanding*. Ithaca: Cornell Univ. Press.

—— (1977). "About Proper Names", *Mind*, 86: 319–32.

ZWICKY, A., and SADOCK, J. (1975). "Ambiguity Tests and How to Fail Them", in J. P. Kimball (ed.), *Syntax and Semantics 4*. New York: Academic Press. Pp. 1–35.

—— (1984). "A Reply to Martin on Ambiguity", *Journal of Semantics*, 3: 249–56.

Index